Straight Up
A History of Vertical Flight

Steve Markman
&
Bill Holder

Schiffer Military/Aviation History
Atglen, PA

Acknolwedgments

I. Claude Morse, AEDC Public Affairs
2. Mike Tull, Boeing Public Relations
3. Alex Lutz, San Diego Aerospace Museum
4. Bill Wagner, Ryan Teledyne
5. Dave Menard, Research Department, Air Force Museum
6. Laura Romesburg, Aeronautical Systems Center History Office
7. Frank McGrane, U.S. Army Transportation Museum
8. Dr. Raymond Puffer, Air Force Flight Test Center History Office
9. Jack Beilman, Niagara Aerospace Museum
10. Darwin Edwards, Museum of Aviation, Warner-Robins Air Force Base GA
11. Tim Edwards, U.S. Army Aviation Museum, Ft. Rucker AL
12. Mark Brislawn, Boeing Aircraft, Long Beach
13. Colby O. Nix, Bell Helicopter Textron
14. Bill C. Martin, Bell Helicopter Textron
15. Jerry Pickard, Bell Helicopter Textron
16. Dr. Larisa Zaichik, CentralAeroHydrodynamics Institute, Zhukovsky, Russia
17. Tatiana Belyaeva, Moscow, Russia
18. Bernie Lindenbaum, Dayton Ohio
19. Bill Blake, Air Force Research Laboratory
20. Rodney Clark, Air Force Research Laboratory
21. Russ Osborn, Air Force Research Laboratory
22. Dr. Dave Moorehouse, Air Force Research Laboratory
23. Dr. Gerry Gregorek, Ohio State University
24. Paul Nann, Darlington, England
25. Short Brothers and Harland, Belfast, North Ireland
26. Ulrich Seibicke, German Air Force

Book Design by Ian Robertson.

Printed in China.
ISBN: 0-7643-1204-9

We are interested in hearing from authors with book ideas on related topics.

Published by Schiffer Publishing Ltd.
4880 Lower Valley Road
Atglen, PA 19310
Phone: (610) 593-1777
FAX: (610) 593-2002
E-mail: Schifferbk@aol.com.
Visit our web site at: www.schifferbooks.com
Please write for a free catalog.
This book may be purchased from the publisher.
Please include $3.95 postage.
Try your bookstore first.

In Europe, Schiffer books are distributed by:
Bushwood Books
6 Marksbury Avenue
Kew Gardens
Surrey TW9 4JF
England
Phone: 44 (0) 20 8392-8585
FAX: 44 (0) 20 8392-9876
E-mail: Bushwd@aol.com.
Free postage in the UK. Europe: air mail at cost.
Try your bookstore first.

Contents

Chapter 1   Introduction 4

Chapter 2   Compound(Unloaded Rotor) VTOL Systems 11

Chapter 3   Tilt-Wing VTOL Systems 22

Chapter 4   Tilt-Thrust VTOL Systems 38

Chapter 5   Vectored Thrust VTOL Systems 71

Chapter 6   Lift-Cruise VTOL Systems 83

Chapter 7   Lift-Fan Thrust Augmentation VTOL Systems 109

Chapter 8   Deflected Slipstream VTOL Systems 116

Chapter 9   Ejector Augmented Thrust VTOL Systems 122

Chapter 10   Tail-Sitter VTOL Systems 129

Chapter 11   Zero-Length-Launch Technique Systems 141

Chapter 12   Vertical Flight Maneuverability 152

Chapter 13   VTOL Concepts That Did Not Make It 161

Chapter 14   Almost VTOL Aircraft Systems 168

1

Introduction

The growth of aviation certainly parallels the history of the twentieth century. It has shrunk the size of the world, putting virtually any destination on the planet no more than twenty-four hours away. Aviation has also changed the way military strategy is planned and wars are fought. But herein lies one of aviation's limits, a problem recognized years before the Wright Brothers made their first flight.

An aircraft has to achieve flying speed before being able to leap into the air, and it must slow down and stop in a controlled manner after losing flying speed. This created the need for runways on which to accelerate and then on which to stop. Freeing aircraft from this need for runways always was one of the goals of aircraft designers. Being able to take an air vehicle straight up always had a magic attraction, and it's no different today.

To an extent, the development of the helicopter accomplished this objective. It allowed for vertical take-offs and landings from unprepared sites, as well as vertical climbs and descents. The helicopter even could land safely under most conditions following an engine failure. But helicopters present their own set of limitations. They are complex, expensive to maintain, have short range, limited payload, and due to aerodynamic intricacies of the main rotor, have a limited top speed.

Even the best performing helicopters barely exceed a 200 knot speed or a 300 mile range. While the helicopter is a very versatile vehicle, these shortcomings limit its effectiveness to a very restricted segment of the overall aviation spectrum.

The ideal solution was always believed to be some type of vehicle that would possess the agility to take off and land like a helicopter, yet fly at the higher speeds and carry the greater loads of a conventional, fixed-wing aircraft. The earliest attempts naturally were to try variations on the helicopter. The compound helicopter was one approach. By adding a small wing and a propulsion system for forward flight, the rotor would no longer be needed to produce lift, thus eliminating the rotor loading problem. Despite great hope, none ever went into service.

Thus was born the concept of the Vertical Take Off and Landing, or VTOL, aircraft. In general, most were more airplane-like but possessed systems that allowed them to take off vertically. In the years since the end of World War II, the activity to develop VTOL aircraft has been continuous, and during certain time periods, almost frantic. The 1950s and 1960s were the "heyday" of such development. Many prototypes of various designs were flown and a lot was learned. The designs that were tried produced varying amounts of

A VTOL design of the future shows a downtown commuter craft. (Sikorsky Photo)

The concept of a VTOL fighter has long been considered, as this 1960s drawing shows. (General Dynamics Photo)

An early concept drawing of a carrier-based VTOL transport. (General Dynamics Photo)

success and involved just about every possible combination of propulsion system variations and creative uses of aerodynamics. A few things proved to work a little, and a lot of things proved not to work very well at all. It's hard to say which technique proved to be the best. The success or failure of a design didn't depend solely on technology: mission requirements, politics, and corporate goals often affected decisions just as much. Despite all this work that was done, the fact is that by the 1970s, with only the Harrier and the Yak 38 military jets in service, and the V-22 Osprey about to enter service in the late 1990s, there still remains a lot of research that needs to be performed before VTOL aircraft will be more widely used across the aviation spectrum.

In addition to the strict interpretation of vertical take-off and landing, this book takes a look at other designers' concepts in related areas: short take-off and landing aircraft (not eliminating runways, but allowing shorter ones), rocket assisted take-off (to at least get in the air without a runway), and all attitude maneuvering (nothing to do with runways, but allows the aircraft to be pointed in any position, despite the wing being stalled).

Many aircraft that are capable of taking off and landing in very short distances have been built and have gone into service. Referred to as Short Takeoff and Landing (STOL) aircraft, they've proven successful for civil, general aviation, and military roles. In fact, the VTOL aircraft in service most often operate in the STOL mode rather than VTOL mode because they can carry a greater payload over a longer range. A few of the more interesting STOL programs in recent years are discussed.

Unmanned aircraft used as interceptors and long range bombers were in service by the early 1950s and routinely were launched using rocket engines. In some of these applications of the concept, the rocket booster providing the vertical/or near vertical power was jettisoned. In other applications, such as the Bomarc air defense missile, the solid rocket motor is an integral part of the fuselage structure and was carried along throughout the mission. A logical development was to adapt this technique to manned aircraft. Zero Length Launch was demonstrated during the 1950s using manned fighter aircraft.

Although all discussion to this point addressed achieving vertical flight with the intention of minimizing the need for runways, another dream of aircraft designers was the ability to maneuver the aircraft to a vertical attitude while maintaining full control. Under normal conditions, the aircraft wing will stall at some angle of attack, and attempted maneuvering beyond that point often results in a deadly spin. The onset of wing stall indicated the limit of turning performance. "Post stall" maneuvering became a concept by which the pilot could point his aircraft's nose anywhere, even straight up or to the rear, while continuing to travel forward. While not a maneuver that

The XV-15 shows one concept of VTOL performance with a Tilt-Thrust capability. (US Navy Photo)

Through the years, there have been many VTOL concepts that looked promising at the start, but never made it. Some were very interesting to say the least, but never progressed past the mockup, or even the detailed design stage. The final chapter examines some of their hows and whys.

The purpose, therefore, of this book is to survey the technology of all the many VTOL vehicles, providing an overview of the programs. Being able to fly straight up, to free the pilot from conventional runways and attitude limits, always was a goal of aircraft designers. *Straight Up* takes these designers' dreams and documents their development. Hundreds of reports and technical papers were found and reviewed, as well as tracking down engineers and designers who worked in VTOL and post stall maneuvering programs. In researching day-to-day technical developments, many of which were forty to fifty years old, it was inevitable that conflicts in the data would exist. There are many reasons for this, such as predictions being taken for test results, mixing knots with miles per hour, and outright typos in the source data. In many cases, people's recollections even change with time. There were also occasional contradictions within a report! The authors have attempted to resolve these conflicts, but where they could not, the contradictory information was included.

would be enjoyed by the average airline passenger, the military application of such a capability was obvious. Post stall maneuvering allows the pilot to maintain precise control of the aircraft without entering a spin even though the wing is stalled. To date, there has been considerable investigation into achieving this capability. Such a capability can provide a tremendous advantage in air-to-air combat.

Technical Challenges
Hovering Flight and Transition Between Hovering and Conventional Flight

Hovering Flight

Designing a successful VTOL aircraft requires the designer to overcome numerous problems that do not exist for conventional aircraft. While successful VTOL aircraft have been powered both by propellers and jets, there are three major

requirements for VTOL aircraft that do not exist for conventional aircraft. The sustaining force needed to hold the aircraft in the air, (1) must be able to be directed straight down, (2) must be great enough to overcome the aircraft's weight, and (3) must be able to be directed so that the pilot can maintain attitude control.

The Harrier VTOL has long been an operational fighter with both the United States and Great Britain. It uses a Vectored Thrust concept. (US Navy Photo)

To get a better feel for the obstacles involved, let's take a brief look at the physics involved in the propulsion problem. First, the thrust produced by a propulsive system is equal to the product of the mass flow rate of the air being moved (including the exhaust gas) and the velocity at which it moves. Thus, to produce a given amount of thrust, either a large mass of air and exhaust must be moved at a relatively low velocity (e.g., by a rotor), or a smaller mass must be moved at a relatively high velocity (e.g., a jet). A prop used for hovering flight falls somewhere between these two extremes. Now, let's look at the power required for the engine to produce a specific amount of thrust. It varies with the square of the exhaust velocity. Thus, a jet engine, with its higher exhaust velocity, requires more power to produce a given amount of thrust, as compared to a prop or a rotor. More power means a larger, heavier engine, and higher fuel consumption. To deliver a given payload some given distance, a helicopter will use about 1/4 as much fuel as a propeller VTOL airplane, and only about 1/25 as much as a turbojet VTOL aircraft. It should be fairly obvious that a helicopter is more efficient for missions requiring a large amount of hovering, but their limited forward speed makes them less efficient for longer missions. Jet VTOL aircraft, like the Harrier, must limit their time hovering in order to have an acceptable mission range. In operational use, they normally will make conventional take-offs and landings, performing vertical take-offs and landings only when required by a specific mission or training requirement.

Since the power required for hovering and cruise flight differs greatly, this introduces other problems that the designer must contend with and resolve. Sizing the jet engine for hovering flight means it may have much more power available than it needs for conventional flight (despite the opinion of many that a pilot can never have too much power). Jet engines tend to be more efficient at the higher thrust settings. In conventional flight, the engine may be operating in a lower thrust range where it is less efficient. Some jet VTOL aircraft used engines sized more for the cruise condition, but augmented them for hovering flight by adding smaller, vertically-mounted engines to provide additional thrust only when needed. This leads to better fuel efficiency, but results in a larger fuselage, additional weight, and increased complexity that is used only during a very small part of the mission.

It would appear obvious at first that any VTOL fighter, with more thrust than weight, should be capable of supersonic flight. Aircraft like the Harrier prove this is not always so. One of the reasons is the engine inlet. Jet engines produce more power as the aircraft increases in speed because of ram air being forced into the inlet. A small inlet can deliver plenty of air to the engine at high speed, but inadequate amounts at the start of the take-off roll. The inlet design is a compromise between the needs of high speed flight and an acceptable runway length. In the Harrier, the inlet was sized to draw in enough air at zero airspeed. With the very large inlet, the Harrier pays a significant drag penalty at higher speeds.

Additional problems include stability and control, safety in the event of an engine failure, slipstream impingement on the aircraft, recirculation of exhaust gases, and noise.

Precise control in three linear directions and in three attitudes must be maintained in hovering flight for positioning over a small landing area, then for achieving a smooth touchdown. Conventional control surfaces are useless because there is no air flowing over them. Control must be achieved by some means of directing the thrust or slipstream to control the pitch, roll, and yaw motions and translations fore/aft, sideways, and up. Control surfaces can provide some effect only if there is air flowing over them. Lacking such use by the control surface, other techniques must be utilized. The most commonly used one is ducting some of the bleed air out to the wing tips, nose, and tail to control all translations and angular motions. Other techniques include taking advantage of the direction of rotor rotation so that engine torque can produce yaw, and differential thrust if engines or props are at or near wing tips.

VTOL aircraft have no inherent static or dynamic stability when hovering. In conventional flight, the location of the lifting surfaces insures that the aircraft will return to its original position when disturbed, and that any oscillatory motion will damp out by itself, or at least be of a long enough cycle time that it will not be too bothersome to the pilot. As an example, consider what happens in conventional flight when a wind gust makes the aircraft pitch up. The lift on the horizontal tail changes, producing a pitching moment that tends to restore the aircraft to its original attitude. Keep in mind that this does not just happen…the designer had to choose a location, shape, and size for all components to make it happen. In hovering flight, there is no restoring force to return the aircraft to its original position when disturbed; the forces produced by the engines do not change relative to the body of

The XV-4B VTOL fighter used the Lift-Cruise technique. (USAF Photo)

Although not providing pure VTOL capabilities, the YC-15—using the Deflected Slipstream concept—demonstrated excellent STOL capabilities. (McDonnell Douglas Photo)

This strange little craft, the Vertijet, was a pure Tail-Sitter VTOL fighter, one of many concepts that didn't make it to operational service. (Air Force Museum Photo)

the aircraft. Thus, if disturbed in pitch (nose up, for example), the hovering aircraft will travel backward and an immediate restoring aerodynamic force will not be produced to return the aircraft to level flight. The critical element here is *immediate* restoring forces…as the aircraft builds speed, forces will develop, but this will happen much too slowly for stability purposes. Although unstable aircraft can and have been flown, the workload is extremely high. It is very easy for the pilot to lose control, not a very desirable situation when only a few feet off the ground. Thus, it is highly desirable to have some type of artificial stability system that senses the aircraft's attitude and automatically adjusts the thrust direction immediately to produce these needed restoring forces. The pilot flying the aircraft is unaware of these small, but continuous inputs. To him, the aircraft feels stable and his workload is at an acceptable level.

Engine failure during hovering is another obvious critical problem. The design should be such that a failure would allow the pilot to maintain control and make a safe landing. For single engine aircraft, there is no choice but to descend. Helicopters and tilt rotor VTOL aircraft can autorotate safely to the ground. However, jet aircraft such as the Harrier will fall to the ground. Multi engine jet aircraft usually were designed so that they could still hover safely with one engine failed. Multi engine tilt wing, ducted fan, and tilt rotor aircraft needed to have their props interconnected so that all would continue turning at the same speed if an engine failed. Some VTOL aircraft built solely for research purposes lacked such a fail safe capability.

Slipstreams and jet exhaust from a hovering aircraft pose many problems. They can erode, or even melt, a runway, or blast a depression into a soft surface. Recirculation is a phenomenon in which the hot exhaust hits the ground, spreads out, and rises (because it is hot). At some point, the engine draws it back toward the aircraft, pulling it into the engines and/or making it flow over the aerodynamic surfaces. Aerodynamic effects can be either positive or negative, depend-

ing on the unique shape of the airframe and location of the engines. The engines lose power when they draw in the heated air. In addition, loose particles of sand or stone can be drawn into the recirculating flow and damage the engine. Even helicopters, with their relatively low slipstream velocities, will kick up a considerable amount of dust if they hover over one spot of bare ground for too long.

Another recirculation effect occurs particularly on aircraft with props or engines near the wing tips. As the flow travels along the ground, the portion that travels inward toward the fuselage will meet the inward flow from the other engine. When this happens, most of the combined flow travels upward toward the bottom of the aircraft. This is referred to as the "fountain effect." Three things now can happen. First, the upward flow can travel around the fuselage and produce aerodynamic lift in the downward direction, a phenomenon referred to as "suckdown." Second, the upward flow hitting the bottom can push the aircraft up, usually a desirable effect in that it acts as a cushion. This makes the design of the bottom of the fuselage very important. If the shape does not lend itself to taking advantage of the fountain effect, then some things can be done to help. The most common is to add sets of strakes on the bottom of the fuselage to help trap the fountain and prevent it from moving up the side of the fuselage. The third effect of the fountain is that it can suck in sand or stones and blast them against the bottom of the fuselage and wing.

During transition, when a considerable amount of thrust is still being directed downward, the exhaust can pull in ambient air, altering the way it flows over the wing. Usually this will decrease the lift produced because it decreases the angle of attack.

Hovering can also be dangerous for personnel on the ground. Hot, high speed exhaust can throw up dirt and other debris and even blow over ground support equipment, causing damage and injury. Any or all of these phenomena can affect any given aircraft, and the overall effect can be either positive or negative.

The 1960s-vintage Matador surface-to-surface missile used the Rocket Assist Zero-Length launch technique. (Bill Holder Photo)

Recirculation can affect the stability and control characteristics in hover. If the recirculating flow still has substantial speed as it hits the aircraft surfaces, it can produce unpredictable forces. This can be complicated even more if the flow reflects off rough, uneven, ground or nearby buildings, and if there is turbulence and gusty winds.

Hovering near the ground can affect the thrust in other, more obscure ways. As the slipstream or jet exhaust hits the ground, it spreads out and slows down. This causes an increase in pressure within the flow. This higher pressure air can act as a cushion. For a helicopter, this has the beneficial effect of increasing lift as much as forty percent for a given power setting. The dominant factor in this phenomenon is the ratio of the rotor diameter to the height of the rotor from the ground. The closer the rotor can get to the ground, the greater the increase in efficiency. Near the ground the dynamics of the air flowing around the tip of each rotor blade changes, just as the drag of a fixed wing aircraft decreases as it descends into ground effect. For a typical propeller VTOL aircraft, this ratio will be lower than for a helicopter and much less benefit will be realized. On the other hand, a ducted fan or lift fan will loose lift near the ground because the higher pressure on the back side of the fan tends to make the blades stall. This is because there is no change in the dynamics of the air flowing over the blade tip, since the tip is very near the duct wall. For the jet VTOL there is a slight loss in thrust because of the increased back pressure. However, this usually is very small because the exhaust pipe needs to be kept as far from the ground as possible to minimize ground erosion. Thus, the ratio of the exhaust pipe diameter to the height from the ground is usually small. It can be further minimized by lengthening the landing or by having the aircraft operate from a metal grating several feet above the ground.

VTOL aircraft tend to be noisier than their fixed-wing counterparts. This is not unexpected because they require much greater power in order to hover. The noise of propeller VTOL aircraft can be reduced by increasing the blade area (which decreases the blade loading) and reducing the Mach number at the blade tip (by either a lower RPM or smaller diameter). For ducted fans and lift fans, the number and spacing of the fan blades and stator vanes are important factors in determining the noise level. For the turbojet VTOL, there is little to do to cut the noise. However, VTOL aircraft can climb either straight up or at a very steep climb angle. This may keep most of the noise directly over the airport, minimizing the disturbance to airport neighbors.

Loss of lift during transition - The transition between hover and cruise flight can involve many aerodynamic, mechanical, and stability and control phenomena all occurring at the same time. It is a busy task for the pilot, requiring full concentration and even an extra hand or two. Aircraft with different VTOL concepts will place different specific demands on the pilot, but the following will generally occur. In hover, the engines generate all of the lift force and the wing generates none. Virtually all of the thrust points straight down. To start the transition, the pilot directs some of that thrust to the rear so the aircraft starts moving forward. This can be done by tilting the wing, redirecting some of the engine exhaust rearward, or tilting the engines or props. As soon as this tilting occurs, a small amount of lifting force is lost by producing the accelerating force. The aircraft may settle unless the pilot adjusts the throttles to reestablish the total lifting force. As the aircraft gains speed, the wing starts generating lift, and the pilot again must adjust the throttles and tilt angles or thrust deflectors to maintain the desired climb angle, attitude, and airspeed. The pilot continuously is adjusting the amount of lift provided by the engines and by the wing until the transition is complete and the wing sustains the aircraft's entire weight.

Having a maneuver capability at attitudes approaching vertical has long been a desire for advanced fighters. In recent years, much research for that capability has been attempted. (Rockwell International Photo)

Some strange VTOL military vehicles have evolved through the years, such as this flying jeep. (Air Force Museum Photo)

Transition Flight

The transition range usually is defined as the speeds between hover and the minimum speed at which the aircraft can sustain its weight solely from aerodynamic lift. VTOL aircraft usually are classified according to the propulsion system concept they use for performing the vertical take off, hover, and transition. Other authors have grouped the aircraft differently, especially the hybrid designs that combine different propulsive concepts. The authors feel this is an academic point that has nothing to do with technical merits and chose to classify the aircraft according to five basic categories, plus sub categories and combinations:

Aircraft Tilting - The entire aircraft tilts from the vertical position for take-off and landing to horizontal for normal flight. These include the "Tail-Sitters," such as the Ryan X-13. While several of these types flew successfully, the pilot lied on his back during take-off and landing, requiring him to look over his shoulder to position the aircraft for touchdown.

Thrust Tilting - Three different concepts tilt the thrust: the Tilt Wing, such as the LTV XV-142, the Tilt Engine/Prop, such as the Bell Air Test Vehicle and XV-15, and the Tilt Duct, such as the X-22.

Thrust Deflection - Two different concepts are used. Thrust can be deflected after it leaves the engine or prop, usually by means of an extremely large flap, as was done by the Ryan VZ-3. Thrust also can be deflected within the tailpipe and nozzle assembly to produce Thrust Vectoring, such as with the Harrier.

Thrust Augmentation - Two different thrust augmentation techniques have been used with varying amounts of success. Aircraft such as the XV-4A blew engine exhaust into chambers in such a way that it drew ambient air into the flow to produce a net thrust increase. Aircraft such as the XV-5 used the jet exhaust to drive fans that drew in ambient air to increase the thrust. While ducted fans such as the X-22 also produced thrust augmentation, the authors chose to include them as their own Tilt Duct concept.

Dual Propulsion - This involves the use of separate engines for lift and for forward propulsion. Some aircraft have been purely dual propulsion, such as the Short Brothers' S.C.1. This concept is referred to as Lift+Cruise. Others, such as the XV-4B, not only had separate lift engines, but the cruise engine could double as a lift engine by deflecting the cruise engine's thrust downward. This hybrid configuration is referred to as Lift + Lift/Cruise propulsion.

Test pilots vs. line pilots

Test pilots are not supermen, just very highly trained, and then specially trained for any particular test program. They must understand the system they are testing as well as the designer understands it. They must be able to establish the test condition and perform each test exactly as the designer demands. They must observe the aircraft response and sort out the multitude of things that will happen simultaneously. Finally, they must be able to describe precisely what they experience in flight to the designers. The goal of flight testing is to insure that an aircraft is safe and manageable by the average pilot. If special techniques are needed, it is up to the test pilot to develop them and determine if the average pilot can master them in a reasonable time. The fact that so many of the VTOL research aircraft from the 1950s and 1960s crashed is testament to the fact that even these specially trained pilots could not maintain control under some extreme conditions or system failures.

2
The Compound (Unloaded-Rotor) VTOL

With many types of VTOL concepts, there are very fine differences between the different techniques. There will be no discussion on the "pure helicopter-type" of VTOL, but its first cousin, the Compound (or Unloaded-Rotor Convertiplane) will be discussed in this chapter.

One of the biggest constraints faced by the helicopter through its evolution has been the lack of forward speed the design possesses. The Compound concept alleviates that problem with the addition of standard propulsion which is brought on line once the vertical portion of the take-off is accomplished.

The attractive concept, though, doesn't come without several penalties, namely increased cost, certain instability problems during the transition period, and the expected greater complexity with the additional propulsion system onboard.

Then, there is the consideration of ending the operation of the helicopter-style blades once their function has been accomplished. There have been considerations of stopping the overhead blades in flight, and even the more interesting idea of stowing the rotor (in some cases actually flushing the blades next to the fuselage), which would allow the Compound vehicle to acquire greater speed without the speed limitations of the rotor.

Such a concept was the so-called Canard Rotor/Wing system, a McDonnell Douglas concept where the exhaust and bypass gasses were ducted to nozzles near the rotor/wing tips to maintain rotation. For fixed-wing flight, the rotor/wing was stopped and locked at 120-150 knots. The gas was then diverted to provide conventional forward thrust.

In certain Compound concepts, the rotor was allowed to free-wheel, but there is very little lift provided. Once the conventional engines have taken over for horizontal flight, most of the aircraft weight is supported by the wings. Since the rotor in this concept is not under power during horizontal flight, the rotation rate is much lower, which tends to lower the drag produced from the rotor. It is this increased rotor blade drag, when the rotor is operating at high speeds in a pure helicopter application, that prohibits the speed of the helicopter.

The Compound concept was initially investigated in 1951 when McDonnell was awarded the contract to develop a Marine Assault Transport. Called the XHRH-1, the project would never move beyond the prototype stage. The design featured conventional propellers and stub wings which provided all thrust and lift for conventional horizontal flight.

During the same time period, McDonnell also considered other Compound concepts, including the Model 113, which was offered as a medium cargo-and-troop transport. It would have had a pressure-jet-driven rotor, but versions with a shaft-driven rotor were also considered. Various powerplant combinations were also studied. For vertical flight, the turboprop powerplants were to drive a compressor which pumped high pressure air out to the tips of the rotor blades where it was mixed with fuel and burned to provide tip-jet power for the rotor. The craft was to have weighed about 15 tons and be capable of carrying 30 troops at 220 knots.

In 1962, a Lockheed XH-56 helicopter was modified into a Compound Helicopter configuration. The modification involved the installation of a 2,600 pound thrust J60 turbojet

Gerard Herrick in front of his "Convertoplane" in 1937 at its first public demonstration. (USAF Photo)

The "Convertoplane" in 1937 at its first public demonstration. (USAF Photo)

mounted on the port side of the fuselage. Small wings were also added on the fuselage.

In 1972, the Compound concept continued with a joint agency program between NASA and the US Army Mobility Command, requesting submissions for the design, construction, flight test and 1976 delivery of the so-called Rotor Systems Research Aircraft (RSRA). The program would never advance to the metal-cutting stage.

But there were a number of systems that did make it into the air, such as the McDonnell XV-1, Soviet Kamov Hoop, the British Fairey Rotodyne, the French Sud-Ouest SO 1310, and the German VFW H3-E Sprinter—all of which are addressed in detail by this chapter. Note the foreign involvement in development of Compound Helicopters.

It came sooner than later with the announcement in mid-1998 of a stopped-rotor vehicle sponsored by DARPA and Boeing. The program was known as the Canard Rotor Wing (CRW) and included a large front canard and twin boom horizontal tail, and a pair of overhead rotor blades with power coming from a turbofan engine.

Another view of the "Convertoplane." (USAF Photo)

The vertical take-off would be accomplished with the rotor blades, through a reaction-powered system, with energy provided by the propulsion system. The transition would occur when enough forward speed was acquired to allow the lift to be accomplished by the wing surfaces, at which time the rotor would be stored in a fixed position and become another wing surface.

Late in that same year, another Compound system was approved, this time the development of an unmanned aerial vehicle (UAV) which incorporated a canard rotor-wing and a total VTOL capability. The Dragonfly, designed and developed by Boeing, could find application with operation from ship decks.

An extremely small craft, the Dragonfly has a 17-foot fuselage length with a forward set of canards with an 8.5 foot span.

The Dragonfly is designed to lift-off, hover, and transition into horizontal flight using its two-bladed rotor. As horizontal flight is achieved, the canard and tail surfaces provide lift, unloading the rotor and allowing it to eventually be stopped, locked into position, and then serve as a conventional wing configuration.

The old Compound concept that has been tried through the decades certainly isn't dead, and continues to be an attractive VTOL technique into the next milleniam.

McDONNELL XV-1 UNLOADED ROTOR CONVERTIPLANE (U.S.A.)

System: Built-from-scratch prototype
Manufacturer: McDonnell Aircraft Corporation
Sponsor: USAF, US Army
Operating Period: 1949-1956
Mission: To test the concept and determine the feasibility of an unloaded-rotor principle for possible application for larger aircraft.
The Story:
Looking at the XV-1, it's easy to see the parts and pieces of both a helicopter and a conventional aircraft. The concept carried both a horizontally-mounted helicopter rotor along with a conventional pusher prop for normal flight. It was the first such VTOL developed in the United States. The US Army was very interested in the system for liaison and observation missions, along with tank and artillery spotting. There were also considerations for using it as an airborne ambulance where it could carry two litter patients and an attendant plus the pilot.

The actual development of the XV-1 took place in conjunction with the Air Force's Wright Air Development Center and the Army Transportation Corps. The XV-1 was one of three VTOL concepts under investigation in 1954. Others included the Bell XV-3 Tilt-Rotor and the Sikorsky S-57 Retractable Rotor vehicles.

The 30-foot-long fuselage was interesting, being pure helicopter in the forward portion and a P-38-style boom arrangement with twin vertical tails (which were about ten feet in height) on the aft portion of the plane. The horizontal stabilizer was mounted between the booms. Another helicopter similarity was the use of a skid instead of a normal landing gear.

The XV-1 had a gross weight of 4,800 pounds with an empty weight of 3,650 pounds. The fuel capacity was only 83 gallons weighing about 500 pounds. The payload was 410 pounds.

There was room for a crew of four or five in the rather large cockpit. With its large glass area in the cockpit area, the XV-1 would have been an observer's dream with visibility available in all directions with the exception of directly below.

Over the wing was a hump surmounted by the rotor head, and aft of which was the engine cowling ending with a spinner.

The propulsion system was interesting, and complicated, in that the 550 horsepower Continental R-975-42 525 horsepower reciprocating engine provided the power to drive both rotors. The top rotor was NOT driven directly by a hookup with an engine shaft, but by pressure-jet units on the rotor tips which were supplied with compressed air ducted through the hub and rotor blades. The arrangement provided the XV-1 with a ceiling of about 11,800 feet and a cruise speed of about 120 miles per hour. The rotor was used effectively only during the lift and descent phases, while in cruising flight it auto-rotated and produced only about 15 percent of the lift. The cruise flight was then supplied in the conventional manner by the pusher propeller located on the aft of the crew compartment and had the capability to push the plane forward at a maximum, and impressive, 200 miles per hour.

The 26-foot-span fixed wing featured a slight sweep on both the leading and trailing edges, although the angle of sweep was greater on the leading edge. It was also designed with a high aspect ratio and joined the fuselage at the same level as the top of the cockpit, directly below the mounting for the upper rotor. The wing terminated in the fuselage into a large bulbous housing on either side of the fuselage.

Basically a new concept, the program moved slowly toward actual transition from vertical-to-horizontal flight. The program began in 1949, but it would take until 1955 to get that first goal achieved. The Air Force then entered the program, where its maximum speed capability was accomplished. Following the Air Force evaluation, all research on the program was discontinued.

In evaluating the program, the consensus was that the basic concept was sound, but the piston engine powerplant could not provide the needed performance to optimize the design. It was felt that use of a gas turbine engine in this application would solve the problem.

Other countries, though, would look at the concept and take it to the next level. But much of the early work had already been accomplished by the XV-1 program which proved that it could be done.

PIASECKI 16H-1 PATHFINDER COMPOUND HELICOPTER (U.S.A.)

System: Use of existing helicopter structure
Manufacturer: Piasecki Aircraft
Sponsor: (Initially a company venture), later US Army/US Navy
Operating Period: 1960s time period
Mission: To develop an eight-place "Ring-Tail" Compound Helicopter
The Story:

The Pathfinder was an interesting Compound Helicopter concept that was carried out during the 1960s by Piasecki Aircraft. Although the project never reached fruition, the concept did show promise during its flight test program. There was interest from both the Army and Navy who shared a joint development contract.

The Compound Pathfinder used a standard helicopter-style overhead rotor which was off-loaded by a small low-mounted fixed wing when the craft transitioned to high-speed horizontal flight. The three-bladed rotor was mounted on a streamlined pylon which was powered by turbine engine. The small tapered low wing was equipped with standard flaps. To the rear of the fuselage were cruciform fins which supported an annulus which contained a propeller. That rear installation performed the function of anti-torque and directional control. The model also mounted a retractable landing gear carriage.

In operation, the Pathfinder took off like a standard helicopter. Acceleration was then achieved by directing power from the rotor into the rear propeller, then as the speed increased, the fixed wings took on a larger share of the lift. It should also be noted that the Pathfinder could also be used in a STOL mode, which greatly increased its payload capability. The reverse transition for landing exactly reversed the take-off process.

There were actually two versions of the Pathfinder, the first of which was the -1 version, which first flew in 1962. The similar Pathfinder II, the 16H-1A, was completed in 1965.

The -1A version was considerably larger than the initial prototype, with a three-foot bigger diameter rotor at 44 feet, and its maximum speed was much faster at 225 miles per hour, compared to 178 with the first version. The gross weight of the IA model was 2,300 pounds heavier than the first model. The two models used different powerplants; the first used a United Aircraft of Canada PT6B engine, while the later version used a General Electric T58 turbine engine.

NAGLER VG-1 VERTIGYRO COMPOUND HELICOPTER VTOL (U.S.A.)

System: Prototype system built from existing aircraft components
Manufacturer: Napier Helicopter Company
Sponsor: Company program
Operating Period: Early 1960s time period
Mission: To develop a combination aircraft/helicopter-type vehicle capable of Compound Helicopter capabilities

The Story:
A little known Compound-type system was developed by the now-gone Napier Helicopter Company in the early 1960s. The craft was a one-of-a-kind system, which certainly could be discerned at first look at the machine.

The prototype was constructed in only six months from a collection of existing aircraft parts and pieces. It used a Piper Colt fuselage and powerplant (a Lycoming four-cylinder 108 horsepower engine), while the over-fuselage rotor system and controls came from a French Sud-Aviation Djinn helicopter.

Another power source existed with a Garrett gas turbine compressor, which provided the compressed gas pressure necessary to power the rotor-tip-mounted propulsion units. A number of new components were incorporated in the design, including the fins, certain control components, and new skinning at portions on the rotor mounting area.

The design of the strange little VTOL then had helicopter controls inserted in place of the standard Colt controls, although the elevator trim control system of the Colt was retained. Yaw control of the Vertigyro during the hovering condition was a complicated technique by movement of the rudder within the turbine exhaust slipstream. The movement was controlled by the pilot using movement of the rudder pedals.

A twist-grip pitch control lever in the cockpit enabled control of the turbine air-bleed valve. Interestingly, the turbine always operated at the same speed.

The company claimed a number of advantages to its unique design, including outstanding operating economics and better performance in either operation as an autogyro or pure helicopter.

The plane's operating technique involved flying the plane in the pure helicopter mode with all the power directed to the rotor with the front propeller feathered, or the pure autogyro condition with all the power being provided by the conventional front propeller.

Initial flight testing of the Vertigyro occurred in January 1964. Both the different flight characteristics of the plane were successfully demonstrated.

The success saw the company start development of a VG-2 version with a more powerful engine. Neither the VG-1, or any of its follow-ons, though, would ever make it to production.

The Avian Model 2/180B. (Avian Photo)

Mock-up of the Marine XHRH-I Compound VTOL. (Boeing Photo)

Model of McDonnell Douglas Compound Model 113 VTOL concept vehicle. (McDonnell Douglas Photo)

AVION MODEL 2/180 CONVERTIPLANE (CANADA)

System: Built-from-scratch prototype
Manufacturer: Avian Industries Inc.
Sponsor: Company program
Operating Period: Early 1960s time period
Mission: To design a small wingless gyroplane.

The Story:

Shortly after the Avian Company was formed in 1959, a company program was initiated to develop an autogyro-type craft which was coined the Model 2/180 Gyroplane.

There were actually two prototypes that were built, the 2/180A which utilized compressed air nozzles on the rotor tips for jump-starts. The 2/180B prototype used a mechanical drive to the rotor.

The prototype began its flight test program in early 1960, but unfortunately was badly damaged in an accident that was not the fault of any design deficiency.

Details of the system included a 180 horsepower Lycoming four cylinder horizontally-opposed air-cooled engine. The engine drives a rear duct-enclosed two-bladed propeller. A belt drive also powered the overhead rotor. The production version used compressed air nozzles at the blade tips for propulsion.

The XH-51N Compound VTOL vehicle. (NASA Photo)

The Lockheed AH-56 Compound Helicopter. (Lockheed Photo)

The Boeing Dragonfly concept Compound VTOL. (Boeing Photo)

The top rotor had three blades, with flapping hinges. The blades had steel tube and leading edge spars, wood core, and fiberglass covering, and used a NACA 0015 configuration.

The plane had a non-retractable tricycle landing gear with a steerable nose wheel and disc brakes.

The craft had a maximum speed of 165 miles per hour and didn't reach stall speed until 25 miles per hour. It had a

Piasecki Pathfinder II. (Piasecki Photo)

Early artist's concept of Pathfinder II. (Piasecki Drawing)

The McDonnell XV-1 was a dramatic step forward in VTOL technology. (McDonnell Photo)

A definite helicopter look was carried by the XV-1. (McDonnell Photo)

The classic combination of conventional aircraft and helicopter—the Nagler Vertigyro. (Nagler Photo)

14,000 service ceiling and a 400 mile range, along with a vertical rate of climb at sea level of 1,000 feet per minute.

Designed to carry two-to-three passengers, the 2/180 had a rotor diameter of 33 feet, a 16 foot, two inch length, and an eight foot, seven inch height. The vehicle weighed only 1,090 pounds empty with a normal loaded weight of 1,720 pounds.

KAMOV K-22 HOOP UNLOADED ROTOR CONVERTIPLANE (SOVIET UNION)

System: Prototype based on an AN-10 fuselage
Manufacturer: Soviet Kamov design organization
Sponsor: Probably Soviet Air Force
Operating Period: Late 1950s to early 1960s time period
Mission: To design a large, functional unloaded-rotor convertiplane with speed capabilities of over 200 miles per hour.
The Story:
The Hoop was an interesting Soviet effort to develop a Convertiplane, resulting in a vehicle that looked amazingly

The Russian Kamov KA-22 was a massive example of the Compound VTOL type. (USAF Photo)

like a conventional aircraft, that is, with the exception of the two wingtip rotors. Otherwise, the fuselage with a single vertical stabilizer and a normal tail assembly made it look quite "normal." A tricycle landing gear featured a forward-fuselage gear along with a spindley-appearing pair of landing gears hanging from the inner wing locations.

The propulsion system, in the form of a pair of 5,600 horsepower Ivchenko turbine engines, provided the propulsive force for both the lifting flat-mounted rotors and the normally-positioned forward-pushing propellers. The tailpipes of these powerplants are assessed to have tail pipes that can be deflected downward to provide additional lift during the vertical flight phase of the trajectory.

Fairey Gyrodyne. (Fairey Photo)

The wingtip-mounted rotors are located some distance above the upper wing surfaces such that they cleared the smaller vertically-mounted propellers. They are estimated to each have a span of about 92 feet.

The entire trailing edge of the high-set cantilever wing is made up of ailerons and flaps, with the rear of the craft fitted with a conventional-type tail.

Making use of the AN-10 transport fuselage design, the Hoop theoretically could have had an impressive passenger capacity of 80 to 100. Also, the fuselage design, with its swept-up lower rear fuselage could incorporate a loading ramp for vehicles and freight.

Little is known about the actual test program of the Hoop as it was designed and developed during the era of secrecy of the Cold War era.

The K-22 demonstrated a significant speed capability, setting a world record for a rotorcraft of over 227 miles per hour in October 1961. That same year, the model was first displayed in the 1961 Soviet Aviation Day display.

FAIREY ROTODYNE UNLOADED-ROTOR VTOL (U.K.)

System: Built-from-scratch prototype
Manufacturer: Fairey Aviation Company, LTD.
Sponsor: British Ministry of Supply
Operating Period: 1950s time period
Mission: To develop an operational unloaded-rotor vehicle for commercial uses with the capability of carrying up to 50 passengers at 200 miles per hour.
The Story:
The initial work accomplished by Fairey with the unloaded-rotor concept was accomplished in the early 1950s with an earlier program, the Gyrodyne, which was powered by a single reciprocating engine. The main emphasis of this development was a commercial application with the design of an intercity rotorcraft.

The emphasis behind the development was the belief that this concept could overcome a number of the inadequacies of standard helicopters, i.e. inadequate payload, insufficient range, mechanical complexity, excessive vibration, and the inability to remain aloft in case of powerplant failure.

Fairey conceived this functional Compound VTOL with commercial applications. The Rotodyne was a giant step forward. (Fairey Photo)

FAIREY ROTODYNE

World's First Vertical Take-Off Airliner

48 Passengers · 185 m.p.h. Cruising Speed
400 Miles Maximum Range

Fairey flaunted its new Rotodyne Convertiplane with this period advertisement. (Fairey Advertisement)

With the Rotodyne concept, a pair of engines each drove a single propeller which provided the propulsion for horizontal flight. The engines also supplied air to a compressor that supplied air to jets that were located on the rotor tips, the power from which provided the hovering capability of the craft. This energy provided aerodynamic lift by whirling the overhead rotor, exactly the same concept as with a standard helicopter.

Once the desired altitude was acquired through this process, the total power of the engines was diverted to driving the propellers. Also, the tip jet operation was ceased and the lift for horizontal flight was obtained by the propeller and the stub wings upon which the engines were mounted. A feature of the design was that adequate single-engine performance was provided following an engine failure at any flight speed.

Transition from vertical-to-horizontal flight, with the rotor autogyrating, was accomplished on April 10, 1958.

The pair of Napier powerplants provided an impressive total of 7,000 horsepower with a gross weight of 39,000 pounds. With a payload of 10,000 pounds, the plane was capable of a range exceeding 400 miles at up to 170 miles per hour. Impressive performance indeed! It's easy to under-

The Farfadet Convertiplane attracted considerable attention in Europe when it was constructed. (Farfadet Photo)

stand why Fairey was so confident that it had a real winner with the Rotodyne.

The technique, though, of the engines serving a dual purpose was the same as the Gyrodyne system, each driving a propeller and providing the compressed air for the pressure-fed wingtip jets. To provide an idea of the size of this machine, vision the fact that the rotor was 90 feet in diameter. For safety aspects, two opposing blades were powered by each engine in case of engine failure. An onboard hydraulic system provided for cyclic pitch control.

The Rotodyne was extremely large, with a cabin volume of 3,300 cubic feet. The logistical attributes of the machine were considerable with rear clam-shell doors allowing the loading of large motor vehicles. A forward-located door permitted simultaneous entry and exit of passengers, which would have allowed a quick turn-around in a commercial airline operation.

It was estimated that a passenger load of as many as 48 could have been carried by the Rotodyne. That passenger compartment was 46 feet long, eight feet wide, and six feet in height.

Like other compound-type VTOL vehicles, had the Rotodyne been minus its trapezoidal rotor-mounting pylon, the vehicle would have looked much like a conventional aircraft. The rotor was mounted at a lofty 22 feet above the ground, providing an idea of the size of the Rotodyne.

The design sported a high wing and a twin-tail configuration. That tail design was interesting in that the lower tail surfaces were oriented straight down, while the upper surfaces were canted at about a 45 degree angle.

The speed capability of the Rotodyne of about 200 miles per hour made it slower than traditional transports, but it could make up for that deficiency with the capability of landing on downtown heliports atop buildings and getting business people to their final destination much quicker. Another attractive aspect of the Rotodyne for this commercial application was the fact that it was projected to have a range of up to 400 miles at gross weight.

The Rotodyne was subjected to a vigorous flight test program of over 350 flights, more than half of them demonstrating 200 hover-to-vertical flight transitions.

The production Rotodyne was to be a somewhat larger vehicle with a rotor diameter of an amazing 90 feet with a gross weight of 30 tons. The propulsion systems would be changed to Rolls-Royce Tyne engines.

There was serious consideration for production of the craft, both domestically and in the United States. That possibility came to light in 1958 when a license agreement was reached between the Kaman Aircraft Corporation (USA) and Fairley. The agreement provided for the manufacture of the aircraft in the U.S. by Kaman.

SUD-OUEST SO 1310 FARFADET UNLOADED ROTOR VTOL (FRANCE)
System: Built-from-scratch Prototype
Manufacturer: SNCA du Sud-Ouest
Sponser: Company program??
Operating Period: Mid-1950s time period
Mission: To develop an effective Unloaded Rotor VTOL aircraft with possible military and commercial applications
The Story:
Interest in the Unloaded Rotor-type Convertiplane VTOL was in place in France during the 1950s and resulted in the construction and testing of the SO 1310 model.

The design of the craft was interesting in that with the exception of the overhead rotor, the plane looked very much like a standard small private plane. It carried a pair of low-mounted stubby wings and a conventional-appearing tail assembly. The plane sat on a tricycle landing gear with the nose tilted slightly upward.

The powerplant for the craft was a Turbomeca Arouste II turbine which was used to power the propeller located on the normal front of the aircraft. The SO 131 is probably the only Convertiplane-Style VTOL which has its forward-pushing—pulling in this application—in that forward position, most being in the rear of the aircraft.

For the overhead rotor, though, there is a completely separate system, a 360 horsepower Arrius II gas generator which

Inflight photo of the SO 1310 in flight. (Farfadet Photo)

fed compressed air for the small combustion chambers at the tips of the rotor, thus providing the propulsive force for the rotor. The overhead rotor is extremely large, and from the photos of the plane that are available, just misses the vertical stabilizer of the tail assembly.

The program had serious goals for the development of the concept with the single prototype being constructed in 1954. An unknown number of flight trials would follow, but that would be the end for the promising program, the exact reason for its demise being unknown.

VFW H3-E SPRINTER COMPOUND HELICOPTER VTOL (GERMANY)

System: Built-from-scratch Prototype
Manufacturer: Vereinigte Flugtechnische Werke (VFW)
Sponsor: Company program
Operating Period: Late-1960s time period
Mission: To develop a Compound Helicopter-style VTOL which exceeded the capabilities of conventional helicopters
The Story:
It all began in 1960 when VFW began research to develop a VTOL concept that exceeded the capabilities of current helicopters. More than a dozen different concepts were investi-

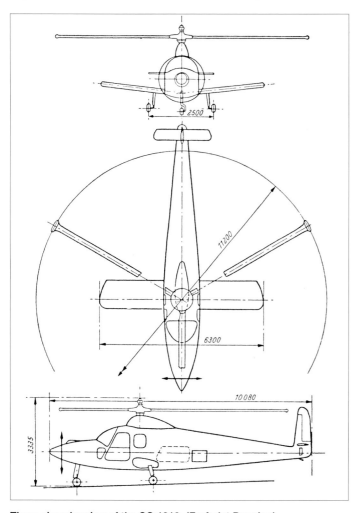

Three view drawing of the SO 1310. (Farfadet Drawing)

gated, and the concept that evolved promising the best growth potential, productivity, speed, and cost effectiveness was the H3-E Compound Helicopter configuration.

The H3-E was built with a mission as a three-seat executive transport, two-stretcher ambulance aircraft, or an agricultural system with a payload of up to 700 pounds.

The design incorporated a compressed air and blade tip-drive rotor. The separate forward-thrust system consisted of fuselage-mounted fans.

Sizewise, the H3-E had a fuselage length of 30 feet, six inches. The maximum height was eight feet, two inches, which included the rotor head. The landing gear track was six feet, seven inches, while the rotor disc area was 640 square feet.

The H3-E was a three-seat craft which had a take-off weight of about 2,100 pounds and an empty weight of 1,100 pounds. The craft provided a payload weight of 590 pounds. The model had the capability of carrying a payload of almost 600 pounds with a fuel load of 460 pounds.

The fuselage was built around an aluminum alloy load-bearing keel which supported the cabin, landing gear, and engine bay structure. The cabin skin was fabricated of a glass-fiber reinforced plastic laminate.

The power unit was an Allison 400 shaft horsepower turbine engine which had a dual purpose. First, it was used to drive a centrifugal compressor in the hover mode. A duct delivered the compressed air through a flexible sleeve to the air distributor around the rotor shaft. Then, the high-pressure air traveled via flexible hose into the roots of the fully-articulated blades.

The overhead rotor consisted of a three-blade configuration, had a diameter of just over 28 feet, and the blades used the NACA 23015 airfoil section. The speed range for the rotor varied from 280 to 480 revolutions per minute with a maximum loading of 3.24 pounds per square foot.

When the air reached the end of each rotor, it was thrust through flush-mounted slot nozzles. A state-of-the-art gearbox contained two bevel gears for the fans and a brake on the compressor shaft for switching the power to the compressor or to the fans. The mechanical layout of the system effectively eliminated the need for conventional transmission and driveshaft systems, hydraulic systems, and a tail rotor.

The technique to achieve near-vertical flight occurred when the rotor was slightly rotated in a standard helicopter style. With the increase in speed, the side-mounted fans were caused to free-wheel within their containing shrouds.

At a certain point in the trajectory, a decision that was made by the pilot, the transformation to full utilization of the fans could be made. Since the fans were already in a windmilling situation, the transition to full fan speed took only about two seconds to accomplish.

Hovering stability was mainly affected by blade hinge offset, blade pitch, angular velocity, disk loading, gross weight and mass moment of inertia of the aircraft.

The performance of the H3-E showed a capability of a maximum cruising speed of 135 knots with the normal cruising speed about five miles per hour less. The maximum vertical rate of climb was carried as 390 feet per minute, with a service ceiling of 13,000 feet.

Early in its program the H3-E underwent a number of test programs. An extensive blade fatigue test attempted to simulate the temperature and pressure cycles inside the blades. The test rig was fully automated, and every five minutes, the temperature and pressure increased and stabilized for 45 seconds before the blade vibrated.

The only two prototypes of the H3 were constructed in 1968, with the first flight test occurring the following year. Before flight, though, there were considerable ground shake tests accomplished. A sophisticated test rig excited the rotor head with a constant force independent of frequency. Ground tests also showed that the vehicle had certain mechanical instabilities at high rotor speeds.

The success of the H3 version resulted in the construction of a follow-on H5 prototype which bore marked similarities to its predecessor. The H5 incorporated many H3 components, including rotor blades (although two feet longer), a modified H3 landing gear, along with H3 controls, tail, and tail boom. The also-similar cabin could be opened with left-side seats capable of folding for litter loading and unloading. Like the H3, there was also no hydraulic system with the H5. There were also a number of Compound Helicopter programs with

VFW called this concept Convertiplane the H3-E Sprinter. (VFW Photo)

H4, H7, and H9 designations. The H4 system actually incorporated a folding rotor. The H9 system was to be a much larger Compound Helicopter system with a 50 foot main rotor diameter and a maximum lift-off weight of approximately 50,000 pounds.

The H7 was to be a new five-seven passenger compound system. Basically a new design, the H7 was to have used some cabin components from the H5, along with modified H5 flying controls.

3
Tilt-Wing VTOL Systems

When you think about the optimum VTOL vehicle, and the technique that logically comes to mind to accomplish it, the idea of a Tilt-Wing vehicle seems so obvious.

To turn the engine thrust in the vertically-down position, just tilt the wing—which obviously mounts the engines—through 90 degrees and you have achieved the condition. The advantage of the technique is that during the transition process, there is optimum flow over the control surfaces during transition, which provides lift for the aircraft. There is also the advantage of minimizing the loss of lift due to the downwash phenomena.

There are, however, certain disadvantages that must be addressed with the Tilt-Wing. When the plane is in the hover position, i.e. ascending to a prescribed altitude with the wing in the vertical position, there are inherent control problems when additional control thrust is sometimes required. Also, it must be noted that in vertical flight, the ailerons perform a different function from their normal roll control function. In this attitude, they are now also performing a yaw control function.

And in a windy weather condition, the control problem with this concept is intensified since the wind is impinging directly on the large surface of the wing, tending to push it with a hard force. Such a phenomena has been coined the "Barn Door Effect."

Once the needed altitude has been attained, it is necessary to make the difficult transition to horizontal flight. At this juncture, some interesting balances must exist. When the wing is coming off the vertical position, there must be an agreement between the rear stabilizer and the wing, i.e. the angle of the wing must be continually matched by the stabilizer as it changes and approaches the horizontal position. With a number of the different concepts researched, there has been a continuing problem of the craft tending to stall with the wing near the vertical position. There is also a requirement for cross-shafting the engines with this concept, since an engine-out situation could be disastrous.

Needless to say, the system for pivoting the wing has to be powerful and dependable. This VTOL technique, of course, adds considerable weight to the craft over that of a conventional horizontal take-off aircraft with that hefty pivoting system.

The initial Tilt-Wing concept occurred with the Converta-Wing concepts from a company founded by D.H. Kaplan. With both Navy and Air Force involvement, the plane had a Tilt-Wing with propulsion on each wingtip. The concept differed from the later Tilt-Wing versions in that the four blades were much shorter and spun at higher speeds, the power provided

Six-engine Tilt-Wing research model mounted in NASA Langley Full-Scale Tunnel. (NASA Photo)

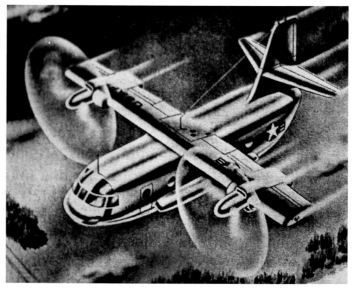

Boeing entry in Tri-Service VTOL Transport Competition. (Boeing Drawing)

North American Aviation entry in Tri-Service Transport Competition. (NAA Drawing)

Chance Vought entry in Tri-Service Transport Competition. (Chance Vought Drawing)

by a pair of Boeing gas turbines. Unfortunately, the vehicle never got a chance to fly, even though it was static tested.

An advanced Tilt-Wing concept evolved as a joint NASA/ Vertol program in the late 1950s that never advanced beyond the wind tunnel stage. A pure Tilt-Wing concept, the craft featured six-propellers driven by a thousand horsepower variable-frequency electric motor. Additional lift was also acquired by double-slotted flaps that covered about 60 percent of the chord. The extremely small-in-diameter propeller mounts gave the plane a completely different look.

The program, and a number of others, greatly advanced Tilt-Wing technology. However, even with all the accomplishments that were demonstrated by the programs, there would be no program that would ever reach an operational status. It appears that the tilting aspect of VTOL research has turned to Tilt-Engine concepts, where a smaller mass has to be rotated through 90 degrees.

And even as the next millenium was approached, there were indications that the old Tilt-Wing concept is not dead. In 1998, for example, Boeing has proposed a new Theater Advanced Transport, called the Super Frog, which would provide near-VTOL performance. The Tilt-Wing system would be reportedly capable of transporting loads two-to-three times the size of the latest C-130J transport.

The numerous Tilt-Wing programs through the years are as follows:

HILLER X-18 PROPELLOPLANE TILT-WING VTOL (U.S.A.)
System: Based on Chase XC-122C and R3Y transports
Manufacturer: Hiller Aircraft Corporation
Sponsor: US Navy, US Air Force
Operating Period: Early 1950s-to-1961
Mission: Develop a testbed aircraft to prove the Tilt-Wing VTOL concept and investigate the problems associated with

Grumman entry in Tri-Service Transport Competition. (Grumman Drawing)

Bell entry in Tri-Service Transport Competition. (Bell Drawing)

the technique. The craft had to be capable of high forward speed in addition to a VTOL capability.

The Story:

Hiller Aircraft was one of the pioneers in VTOL flight and investigated the possibilities of the phenomena following World War II. Its investigations pointed to the advantages to be derived from the Tilt-Wing concept. It generated enough interest from the Navy, with its expertise receiving a contract in 1957 for a four-ton payload, tilt-wing transport.

The company's potential also attracted the interest of the Air Force in 1954. Three years later, the Air Force pushed for the construction of a prototype and a flight test program, and awarded the company a $4 million contract to accomplish the awesome task. Even though the plane was built from scratch as a research-gaining testbed, the X-18 definitely carried the look of an era transport. It would also be the largest VTOL aircraft built to date at the time.

The contract called for a twin-engine, Tilt-Wing convertiplane. In addition to its vertical take-off and landing capabilities, the X-18 was also required to have a significant forward speed capability of about 400 miles per hour.

Another consideration of the flexibility of the plane was that where landing strips were available, the X-18 could be used in a conventional take-off mode, resulting in an increased payload capability. The model was to also be constructed using as much conventional fabrication techniques as possible. The task was aided greatly during its early phases by a series of wind tunnel tests at NACA Langley, Virginia. The first, and only, prototype was completed in 1958, and quickly became involved in an extensive ground test program.

The first actual flight test took place on November 24, 1959. There would be twenty flights in the program, with the program ending in July 1961. It was that final flight of the program that actually spelled disaster.

When a problem occurred with one propeller's pitch control system, it served as the main reason for the termination

The X-18 cockpit had an advanced instrument panel design along with ejection seats. (Fairchild Photo)

of the program. The fact that the engines were not cross-shafted together contributed to worries with the program.

But after the flight testing was over, there was still another important mission to be performed by the X-18. During this phase, the plane would serve to generate a data base for the four-engine XC-142 Tilt-Wing transport that would follow.

During its final days, the X-18 would be damaged when a ground test stand upon which it was mounted failed. Shortly thereafter, the X-18—like a number of the other X planes—would be disassembled and cut up for scrap.

Looking at the mechanicals of the X-18, it quickly comes to light that this was an extremely complicated machine and certainly was pushing the state-of-the-art for the time period.

Vertol/NASA project featured a Tilt-Wing and six propellers. (NASA Photo)

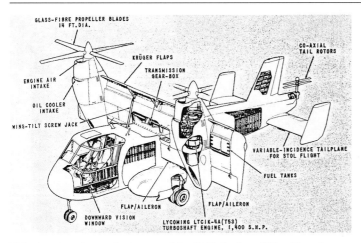

This drawing shows the layout of the X-18 design. (Hiller Photo)

X-18 on Ground Effect Simulation test stand. (USAF Photo)

The propulsion system, although it wasn't immediately evident when viewing the plane, carried three engines. Besides the obvious pair of Allison T40-A-14 turboprops, each mounting a pair of Curtiss-Wright turbo-electric three-bladed propellers and putting out a total of just over 11,000 horsepower, there was also a 3,400-pound-thrust Westinghouse J-34 turbojet located in the aft fuselage that provided the greatly-needed pitch control. It was located in the aft fuselage. The T40, by the way, was the same engine that was used in the Convair XFY-1 Pogo Tail Sitter VTOL fighter.

The thrust from this auxiliary engine was diverted through a pipe that protruded out the rear of the aircraft and terminated with an up-and-down diverter valve. By applying the thrust in either an up-or-down direction, the devise was able to maintain the craft's pitch control.

The T-40 props were huge—16 feet-in-diameter—with the props geared to be counter-rotating. The engines provided significant lift thrust, providing the X-18 with excellent performance through transition from vertical to horizontal flight. Following transition to its forward velocity mode, the X-18 acquired the look of a conventional transport, and it was hard to even identify its VTOL capability.

The cockpit featured a standard cockpit layout, but stability augmentation was built into the roll and pitch axes. Hydraulic boost was used on the ailerons, and a jet diverter and a servo tab boost on the rudder. The only additional control on the cockpit panel was a lever to tilt the wing, which

ABOVE: X-18 lift-off flight sequence. (Air Force Museum Photo)

The Chase VC-122 airframe being converted into the X-18. (Hiller Photo)

X-18 in flight. (Air Force Museum Photo)

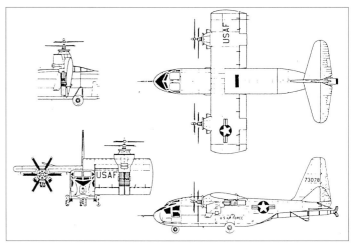

Three-view drawing and photos of the X-18. (Hiller Drawing)

was mechanically locked in the full-down position, but hydraulically locked when in the intermediate positions.

The X-18 airframe certainly wasn't a from-scratch project, but instead was derived from a montage of existing aircraft parts and pieces. The main fuselage, which was 65 feet in length, was a modified Chase XC-122C model with other parts coming from a R3Y transport. Its construction did not require any unconventional techniques or components, which resulted in a great time savings in the overall project.

The fuselage was actually cut in half during the construction process and stretched to a length which was required to meet the center of gravity and landing gear requirements. For simplicity reasons, the tricycle landing gear was fixed in the deployed position. During horizontal flight, the X-18 carried the look of a standard transport, although the wings looked a bit stubby.

The high-set 48-foot-span wing, though, was a new design, with the engines nearly centered between the fuselage centerline and the wingtips. Interestingly, there were no flaps, but ailerons, with the actual tilting of the wing accomplished by a pair of hydraulic pistons. The high-set wing was designed to rotate through a complete 90 degrees, enabling a pure helicopter vertical transition.

The X-18 weighed 27,000 pounds empty and 33,000 pounds loaded. Compared to the other VTOL craft of the time period, the X-18 was a large aircraft and required each main Allison T-40 engine to lift 16,500 pounds, as compared to the 12,000 to 14,000 pounds for each of the Tail-Sitter VTOL projects.

From the pilots' point-of-view, the X-18 was rated as a fairly easy aircraft to learn. That came from the fact that the cockpit controls were practically identical to any turboprop transport of the period, the only addition being the wing-tilt lever.

KAMAN K-16 TILT-WING VTOL SYSTEM (U.S.A.)
System: Based on the JRF Goose Flying Boat
Manufacturer: Grumman with modifications by Kaman
Sponsor: US Navy along with NASA involvement
Operating Period: Early 1960s time period
Mission: Produce a tilt-wing research vehicle
The Story:
In order to possess a Tilt-Wing project of its own, the Navy, after being involved with the X-18 program, decided to go its own way with a Tilt-Wing program. To that end, the service contracted the Kaman Aviation Corporation to design and build

X-18 with wing in full vertical position. (Hiller Photo)

Pitch control is carried out by a single internal turbojet who's exhaust is vented through a long exhaust pipe. (Fairchild Photo)

Artist's concept of the Kaman K-16 Tilt-Wing VTOL. (Kaman Photo)

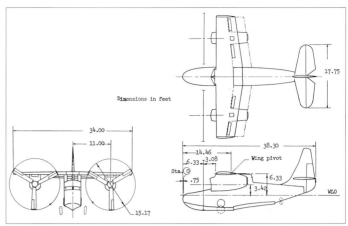

The geometry of the Kamon K-16. (NASA Drawing)

a tilt-wing VTOL vehicle, the program being given the company designation of K-16B.

Like the X-18 program, Kaman turned to existing parts and pieces for the construction of the prototype. It was decided that the fuselage of the Grumman JRF Goose, a flying boat configuration, would exactly serve the purpose for this application, along with the fact that it was already available and wouldn't have to be fabricated from scratch.

The tilting wing, though, would have to be fabricated, which was done in-house. The wing carried a 34-foot span, but unlike other such Tilt-Wing designs which rotated to the full 90 degrees, the Kaman design would only move to a maximum 50-degree position. However, the lifting effect was enhanced since the wing contained large trailing edge flaps that enhanced the downward force effect of the wing when it was in the partially tilted attitude.

Small controllable flaps on the propeller/rotors gave the pilot control of the aircraft at speeds up to 50 miles per hour when the conventional control surfaces were not yet effective. Above 50 miles per hour, the flap control phased out automatically and the conventional controls took over.

The flaps were operated by a cyclic control system so that the propellers could effectively be operated as rotors. The longitudinal cyclic pitch was used to control yaw, while roll was controlled by changes in propeller pitch.

Power was adequate with a pair of General Electric T58-GE-2A turboprops, each driving giant 15-foot diameter propellers, with a projected horizontal speed of up to 300 miles per hour.

The K-16 on a a ground test stand. (NASA Photo)

The K-16 with slat and drooped leading edge mounted in a NASA wind tunnel. (NASA Photo)

K-16 rotor hub details. (NASA Photo)

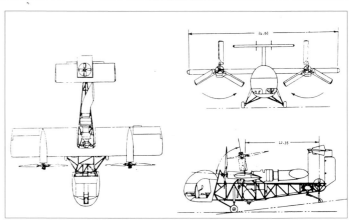

Three-view drawing of the VZ-2. (Vertol Drawing)

When you really think about it, the K-16B really embodied two different VTOL concepts besides the obvious Tilt-Wing arrangement. There was also some contribution from the Deflected Slipstream technique as there was a lift enhancement from the large flaps.

As promising as this aircraft appeared, it would never take to the air. It was, however, tested in the NASA Ames wind tunnel during 1962. The reason it didn't move into a powered flight test stage is not known. Undoubtedly, the number of promising VTOL programs of the time has to play heavily in the decision.

VERTOL 76 VZ-2A TILT-WING VTOL (USA)

System: Built-from-scratch prototype
Manufacturer: Vertol Corporation
Sponsor: U.S. Army, U.S. Navy, NASA
Operating Period: 1956-1965
Mission: To successfully demonstrate the Tilt-Wing VTOL concept with a flyable test bed.

The Story:
Vertol, with its long helicopter history, began involvement with Tilt-Wing investigations in the 1950s with work on its company-designated Model 76 Program. The research would be affirmed with a joint Army/Navy contract, signed on April 15, 1956, with the effort being defined as the VZ-2A Program. With the cost of modern day military contracts, the design and development was for the momentous amount of $850,000!

The resulting vehicle could best be described as "frail looking," with much of the fuselage being of open-tubular construction. If you were to view the vehicle strictly from the front, you would swear that this was a helicopter-type vehicle with its characteristic bubble-type cockpit. That cockpit was located far forward of the wing pivot point and featured side-by-side seating for the two-man crew.

For safety considerations, there were also dual controls which could move control surfaces on the tall-straight-up vertical stabilizer topped with a flat horizontal "T" configuration.

VZ-2 Tilt-Wing VTOL with wing in vertical position. (Vertol Photo)

A rear view of the VZ-2 showing the position of the engine, ie located above the fuselage. (Vertol Photo)

Period Vertol advertisement for the VZ-2. (Vertol Advertisement)

The complexity of the concept was increased with the addition of a pair of ducted fans for pitch and yaw control, both being located in the tail configuration.

Appearing as a last-second add-on, the vehicle's 660 horsepower Lycoming YT53-L-1 turboshaft was mounted by struts above the fuselage. The exhaust was vented outward to the left side of the rear stabilizer. Since the propellers were not attached to actual engines, the units that transferred the power from the fuselage-mounted engine resulted in considerably smaller wing units than the actual engine installations in other Tilt Wing concepts.

A portion of the turbine power was also transmitted through shafting to two ducted fans, one in the vertical and the other in the horizontal stabilizer. These fans, through a pitch-changing mechanism, were used for pitch and yaw control of the craft during hovering and transition flight.

Then, through a complex system which incorporated a cross shaft arrangement, the power was transferred to the pair of wing-mounted rotors which were located close to the center point of each wing. The rotors were extremely large in diameter, at nine and one-half feet in span, and each carried three blades. The variable-pitch rotors, in addition to their primary power requirement, also provided supplemental roll control.

The craft proved to be extremely maneuverable, but was extremely slow with a maximum speed of only 134 miles per hour. For safety purposes, the propellers were interconnected.

Like other Tilt-Wing concepts, the VZ-2A, with a gross weight of only 2,710 pounds, employed a pair of modes of operation during the lift-off to the normal flight phase. As the craft moved away from the hover phase, the normal aerodynamic forces on the plane's wing and tail control surfaces came into play. Also for aerodynamic reasons, the rear fuselage of the plane would later be skinned for smoother air flow.

Dimensionally, the VZ-2A was a small craft with a wingspan of only 25 feet and a fuselage body length of 27 feet. The maximum height was 15 feet.

The VZ-2A was first flown in the late 1950s, however, before that time the configuration had been tested by NASA at its Langley Research Center wind tunnel. The research had shown possible stall problems might be encountered during the transition phase on the initial VZ-2A configuration, which at the time had no wing flaps.

During an early flight test, the VZ-2 wing shows a tilt angle of about 45 degrees during the ascent stage of a lift-off. The aft fuselage would later be closed in. (Vertol Photo)

Vertol VZ-2 in flight during ascent phase. The VZ-2 could be described as spindley at best. (Vertol Photo)

Flat front of the XC-142 cockpit. (Bill Holder Photo)

The first VZ-2 vertical flight took place in April 1957. On July 15, 1958, the first complete transition took place, demonstrating vertical take-off to cruise flight and then back to a vertical landing.

That four-decades-ago flight made VTOL history and proved that the Tilt-Wing concept was viable and certainly surprised many doubters that the technique could work. It was an extremely short flight, and as described by the press as follows:

"The pilot lifted this aircraft as a helicopter, tilted the wings for a short horizontal flight, then converted back to hover and a vertical landing."

The model also served as a great teaching tool, as many pilots were checked out in the unique bird. Once the military was completed with the VZ-2, the plane was turned over to NASA, where it continued to fly until 1965. In all, there were only 450 flights made by this unique craft, including 34 full conversions.

In recognition for its many accomplishments, upon its retirement, the VZ-2A was given a place of honor at the rare bird alcove of the Smithsonian Institution.

LTV-HILLER-RYAN XC-142 TILT-WING VTOL TRANSPORT (U.S.A.)

System: Built-from-scratch prototype
Manufacturer: Ling-Temco-Vought, Hiller, and Ryan
Sponsor: Air Force(Prime), Army, Navy
Time Period: 1959-1967
Mission: A part of the Tri-Service Assault Transport Program, the XC-142 was another attempt to prove the Tilt-Wing concept for a tactical military environment.
The Story:
The XC-142 Tilt-Wing V/STOL program had its roots from the recommendations from a government advisory group in 1959. The group recommended that a full-size aircraft was required, with specific requirements for the Navy and Army. Beyond

One of the XC-142s is on display at the Air Force Museum. (Bill Holder Photo)

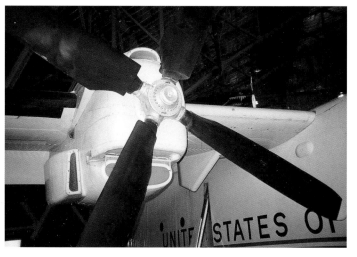

Right inboard engine on the XC-142. (Bill Holder Photo)

Flap and leading edge detail of XC-142 wing. (Bill Holder Photo)

Rear stabilizing rotor on XC-142. (Bill Holder Photo)

the obvious military applications of such a system, there were also many that felt that the plane could also have considerable civilian applications.

The group noted that all previous VTOL programs at the time had been built to illustrate a particular principle, and the fact that it could be accomplished. Few of these concepts had any operational military capabilities. Consequently, with the XC-142, it was decided that this system would be tested in an operational environment.

With that goal in mind, it was decided that the system would fulfill requirements for all three military services. Thus was born the first tri-service VTOL.

In 1961, a Request for Proposal was released, and in September, the proposal from Vought-Hiller-Ryan was announced as the winner. It was also announced that the Air Force would manage the program with the cost of the program to be shared equally by each of the services.

Vought Aeronautics Division of Ling-Temco-Vought was the prime contractor, with Hiller and Ryan serving as the major

subcontractors. Vought subcontracted the design and fabrication of the empennage, aft section, engine nacelles, and wing to Ryan. The overall transmission system and selected components were subcontracted to Hiller, which was also responsible for the flap and aileron fabrication.

Compared to previous test bed experiences, the new V/STOL was a large aircraft with significantly optimistic performance goals, similar to those of the Army's Caribou transport.

The XC-142 grossed out at about 37,500 pounds loaded with an empty weight of about 23,900 pounds. The plane had a fuselage length just exceeding 58 feet, with a maximum height of 26 feet and a sizable wing span of 67.5 feet. The model carried a single tall vertical tail that provided 130 square feet of area. The wings carried large trailing double-slotted flaps the entire length of each wing and were mounted high on the fuselage.

The bulky fuselage was designed to carry significant cargo, with the cargo compartment being 30 feet in length

There was no mistaking the XC-142 with its patriotic paint job—white upper fuselage, blue lower, and red wingtips. (Scott Vadnais Photo)

One of the XC-142s on the roll during the flight test program. (Air Force Museum Photo)

XC-141 with wings and props facing up. (Hiller Photo)

An XC-142 showing its capabilities in an off-road environment. (Hiller Photo)

with a seven foot height and width. That volume equated to about 32 full-loaded troops and gear, or four tons of cargo. In addition to that capability, there was also the ability to carry 1,400 gallons of fuel. There was also a planned capability for auxiliary tanks which would greatly add to the range.

Power consisted of four 3,080 horsepower General Electric T64-GE-1 engines, mounted in nacelles on the wings, which were all cross-linked together. Each drove a giant four-bladed 15.5 foot Hamilton-Standard fiberglass propeller, the tips of each practically overlapping each other.

Later in the program, Hamilton Standard would provide an improved version of the propeller using the 2FF blade design, which featured a wider planform, rounded tips, and a more pronounced twist than the earlier 2EF blades. The goal of the new design was to improve aerodynamic load distribution and overcoming a static load problem.

The four engines also drove a fifth propeller, a three-bladed fiberglass type, in the tail through an interconnected gear and shaft train. Therefore, power was available to turn all five propellers when one, two, or three engines were shut

down. The tail propeller rotated in a horizontal plane and was declutched and braked for cruise flight.

Tying all this power together was obviously an intricate and complicated setup. Through cross-shafting gearboxes, the rotation from each engine was brought together at the top of the fuselage. The power was then sent back to the tail rotor through a tail propeller shaft, into the tail propeller gearbox, and on to the variable pitch tail propeller.

The propulsion system of the XC-142 was definitely an over-powered situation. For example, the plane could lose an engine on take-off and still clear a 50-foot barrier in 400 feet carrying a 10,000 pound payload. Also, with all engines operating, the plane had a rate of climb at sea level of 6,800 feet per minute. On a hot day, even with an engine out, the XC-142 showed a climb rate of 3,500 feet per minute.

Control of the craft during the ascent stage was intricate to say the least with roll controlled by differential propeller pitch. Pitch control was accomplished by the eight-foot, three-bladed variable pitch tail rotor. Yaw control was provided by ailerons powered by propeller slipstream deflection, actually a second VTOL concept being employed in the XC-142.

One of the fleet of XC-142s in the vertical ascent phase. (USAF Photo)

XC-142 in hover mode pulling up a test subject(USAF Photo)

One XC-142 in conventional flight passes another in hovering flight. (USAF Photo)

This hard landing resulted in the destruction of the landing gear system. (USAF Photo)

The craft had a unique capability with the main lift system in that the wing was capable of rotating through 98 degrees instead of the expected straight-vertical position. The wing tilt mechanism consisted of two screw-jack actuators driven by a centrally-located hydraulic motor. The tilt was controlled by a variable rate switch on each collective lever, or by a constant rate switch. This allowed the plane to hover in a stationary mode in a tailwind condition.

The trailing edge of the wings carried three-section, double-slotted flaps in three sections, with the center and outboard sections operated also as ailerons. The flaps were programmed automatically with changing wing tilt, although the pilot had an override capability. Leading edge slats were used for stall suppression, and were mounted outboard of each engine nacelle and operated automatically as a function of flap position. The vertical tail was operated as a standard rudder-and-fin set-up, which supported the slab-type unit horizontal tail assembly.

The magic in the design was probably in the intricate control system, a fully-powered irreversible type with artificial feel forces and powered by dual independent hydraulic systems. Dual cockpit controls, consisting of conventional rudder pedals, control sticks, and collective levers for all take-offs and landings, provided the highest technology of the system.

The XC-142 design also considered logistics implications, in addition to the VTOL design goals, with the tail rotor rigged to fold to the port side to reduce the storage length and protect against damage during a loading operation.

The first XC-142 was rolled out in early 1964 with its first conventional flight being made in September 1964, its first hover three months later, and first transition two months later than that. The Air Force extensively tested the XC-142's ca-

pabilities with cargo flights, cargo, and paratrooper drops, along with desert, mountain, rescue, and carrier operations.

In 1966, one of the XC-142s passed operational tests to prove the model in carrier operations. In quick succession, the plane accomplished 44 short take-offs and landings, along with six vertical take-offs and landings from the USS Bennington.

The carrier trails were accomplished using the number five prototype, which was crewed by both USMC, Navy, and Army pilots. The flight regime covered VTOL operations at a variety of speeds, which occurred at wind conditions from five to 30 knots. A large variety of wings and flap tilt angles were used during the testing. Also, there were landings accomplished with three and six degree glide slopes. In an amazing demonstration, the plane negotiated a 360-degree turn within the width of the flight deck. That same year, one of the prototypes was also tested in an overwater pickup opera-

This accident caused major nose and propeller damage. (USAF Photo)

An XC-142 in carrier trails. (USAF Photo)

tion. The plane lifted a man from a life raft to determine its capability for rescue and recovery. A standard Navy horse collar sling was attached to 125 feet of cable and then lowered through a floor hatch just aft of the cockpit. The tests proved that there were no problems with effects of the propeller downwash or slipstream turbulence.

The program called for the building of five prototypes, but cross-shaft problems, along with some operator errors, resulted in a number of hard landings causing damage to the complete fleet.

The most serious of the mishaps, resulting from a tail rotor driveshaft failure, caused three fatalities. The May 1967 accident took place near the Dallas, Texas, LTV plant and occurred in a heavily-wooded area where fire started after the impact.

The flight plan for the ill-fated prototype included a rapid decrease in altitude from 8,000 to 3,000 feet, effectively simulating a pilot rescue under combat conditions. A nose-over at low altitude followed, from which the crew could not recover. The crash aircraft was XC-142 #1 which had flown 148 times at the time of the crash.

Other incidents included the following:

Aircraft #2-On October 19, 1965, this craft experienced a ground loop causing extensive damage to the wing and propeller.

Aircraft #3-On January 4, 1966, this model made a hard landing in the vertical mode. There was significant damage to the fuselage. The wing of this plane was late mated to the Number #2 for further testing.

Aircraft #4-On January 27, 1966, an engine turbine failure caused the overriding clutch to engage, causing extensive damage to the wing, outboard aileron, the number two nacelle, aft engine shroud, and fuselage. It was later used by NASA for further research.

Aircraft #5-In December of 1966, a ground accident caused major damage to the fuselage, nose, wing, and propellers. The incident was caused by pilot error who failed to activate the hydraulic system, which resulted in no brakes or nose wheel steering.

The final decision on the disposition of the aircraft occurred during the Category II Operational Suitability Program, which was conducted at the Air Force Flight Test Center. The testing consisted of 113 flights, totaling 163.9 hours, which was accomplished between July 1965 and August 1967.

Three of the XC-142s also participated in a major operational test demonstration during the program, where the planes participated in demonstrations of VTOL, STOL, and movement of Jeep-mounted 106mm recoilless rifles, unloading of three-quarter ton trucks with towed 105mm Howitzers, dump trucks, and 1000-pound A-22 containers.

For a typical XC-142 design mission, the plane could operate with a gross weight of 37,500 pounds, including a four-ton payload. At that weight condition, the plane could take off vertically, cruise 230 statute miles near 300 miles per hour, hover for ten minutes, and then land.

One of the limitations found in the plane, even though the overall test results were very positive, was an instability between wing angles of 35 and 80 degrees which was encountered at extremely low altitudes. There were also high side forces which resulted from yaw and weak propeller blade pitch angle controls.

Another XC-142 complaint was the excessive vibration and noise in the cockpit, when coupled with an excessively high pilot workload, and which presented a considerable challenge in the cockpit. The program was a considerable effort, with 39 different pilots flying the prototypes for a total of 420 hours.

The greatest national exposure the XC-142 received during its flight test program occurred when the #4 prototype participated in the 1967 Paris Air Show.

The technology contributions which were derived from the program were felt to have made the program worth its effort. In retrospect, it has to be assumed that if the mechani-

One of the five XC-142s during construction. (USAF Photo)

cal problems experienced with the XC-142 could have been solved, the plane could well have achieved operational status.

The only remaining XC-142, #2, currently is on display at the Air Force Museum at Wright-Patterson Air Force Base near Dayton, Ohio.

CANADAIR CL-84 DYNAVERT TILT-WING VTOL TRANSPORT (CANADA)

System: Built-from-scratch prototypes
Manufacturer: Canadair Ltd.
Sponsor: Canadian Government, Canadair Ltd, CAF, US Navy
Operating Period: 1958-1974
Mission: The purpose of this program was to develop a Tilt-Wing V/STOL transport to serve as a military utility/support role demonstrator.
The Story:
In retrospect, the Canadair CL-84 V/STOL program might have been the first Tilt-Wing concept vehicle to demonstrate considerable success with the technique and generate interest worldwide. The program also generated a great interest from the Canadian military, where the model was given the CX-84 designation.

Interesting is the fact that with this program there was a research and development period (1958-1963) of a considerable length before the first prototype was constructed. The R&D would produce a design that would incorporate a number of significant innovations. Included were the following features:

-A large chord wing which was to be immersed in the propeller slipstream
-Huge propellers on a pair of powerplants

CL-84 with full wing tilt with crewman hanging underneath. (Canadair Photo)

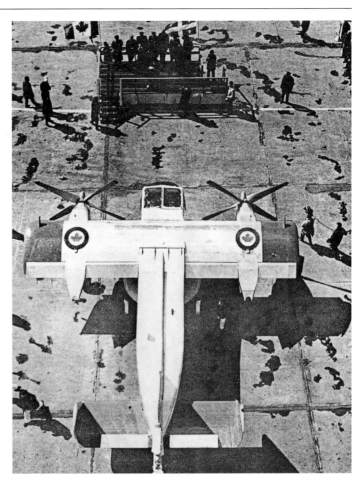

An overhead veiw of the CL-84 shows the extremely far-forward location of the wing. (Canadair Photo)

-All engines, rotors, and the tail rotor which were connected together by intricate shafts and gear boxes
-Conventional-style cockpit controls
-Stability augmentation system for reduction of pilot workload in low-speed flight conditions

Since there was a complete lack of experience with this type of concept, there was considerable scale model testing in Canadian wind tunnels to ensure the design.

Canadair also used its analog computer facility and a cockpit mockup to create a realistic CL-84 flight simulator. The simulator proved to be an effective development tool and was instrumental in the design of the cockpit controls.

The CL-84 was a sizable machine with a maximum height of 14 feet, seven inches, and a rotor tip-to-tip length of over 45 feet. The wings had a total area of 233 square feet with the trailing and leading edge flaps having a sizable 51 square feet. The empty weight of the plane was 7,500 pounds. In a maximum payload configuration, the plane could be loaded to 1,500 pounds for a pure VTOL mission, while 3,600 pounds of payload was accomplishable in a STOL or conventional mission.

Following completion of R&D, the first of four prototypes would be built in two years, with the first vertical flight coming in May 1965, followed by its first conventional flight seven months later.

The first total transition flight was accomplished on January 17, 1966, at the company's Montreal facility. Transition was made from hover to forward flight and back. Flights were made in light snow with wind gusting to 25 miles per hour. The flight, with company pilot W.S. Longhurst at the controls, came seven months ahead of schedule.

The building of that first prototype could best be described as a Lockheed "Skunk Works" operation with word-of-mouth communications being used to supplement the very-sparse drawings during its construction.

The prototypes by normal standards were very small craft weighing only about 8,000 pounds, about a quarter of the weight of the XC-142 that would follow. The wings were only 33 feet in length, and mounted on the underside were a pair of Lycoming T53-LTCIK-4A turboprop engines turning 14-foot-diameter propellers.

The horizontal tail was placed relatively low so that it was below the wing wake during cruising flight and always within the slipstream of the wing-tilt angle. The placement within the slipstream was important in order to prevent abrupt changes in pitching moment as a function of wing-tilt angle.

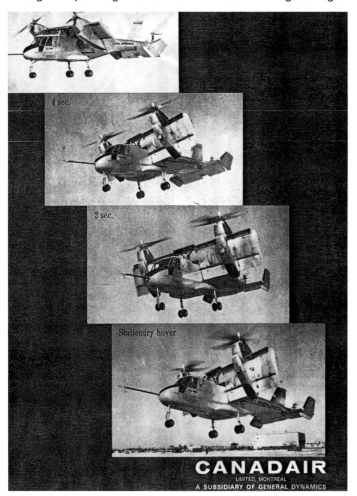

4 sec.

2 sec.

Stationary hover

CANADAIR
LIMITED, MONTREAL
A SUBSIDIARY OF GENERAL DYNAMICS

This period Canadair advertisement flaunts the attributes of the company's CL-84. (Canadair Advertisement)

Therefore, it can be seen that in addition to using the Tilt-Wing concept, the CL-84 also benefited to a lesser degree from the Deflected Slipstream technique.

Pitch control was provided by a pair of horizontally-mounted two-bladed propellers mounted on the rear of the aircraft. When in conventional flight, the props were stopped to minimize drag.

Roll control was maintained by differential thrust from the main engines while ailerons accomplished yaw control.

Capabilities of the CL-84 were impressive, being able to lift 6,500 pounds of fuel and payload in a STOL take-off, or 4,100 pounds of fuel and payload in a VTOL mode. Amazingly, a 35-knot wind could double the VTOL payload capability.

An advantage of the CL-84 came from the pilot's seat where it was piloted pretty much like a conventional aircraft. The pilot sat in the left seat, but a dual set of controls were in place. Even with its complex control mechanisms, the control stick and rudder pedals produced the desired control functions. It really wasn't necessary to learn a completely-new flying technique with the CL-84.

Amazingly, the pilot could fly the plane without actually knowing the wing angle. The innovative engineering of the CL-84 was really brought out by the plane's control system.

A number of significant flying maneuvers were accomplished by this program, including forward flight from hover (wing tilt 88 degrees) to 33 knots (wing tilt 48 degrees) and return to hover mode, demonstration of adequate control in winds gusting to 25 knots, rearward, sideward, and turning flight in and out of ground effect, and sustained flight with hands free of the controls.

A so-called mixing box brought all the control forces together to act as one as it linked the elevators, rudder, ailerons, and propeller blade angles together.

By April 1966, the CL-84 had completed 70 test flights consisting of hover, conventional flight, STOL, and transition flights. Investigation into the high-speed regime followed undercarriage retraction tests, with speeds up to 200 knots in 60-degree banked turns being achieved. Other significant test accomplishments occurred during low-speed maneuvers. For example, the CL-84 easily completed 2G turns at only 90 knots, followed by 200-foot radius turns at 50 knots.

The plane also demonstrated exceptionally stable hovering flights, some with pilot hands and feet free of the controls. The CL-84 was, in fact, hovered and even landed vertically with the stability augmentation system not operating.

The first prototype was flown for two years by 16 pilots for a total of 145 flying hours. A number of military applications were tested, including dropping of external stores, mini-gun firing, simulated rescues from hover, use of a cargo sling, joint operations with a helicopter at seas, and hover downwash tests. A number of United States teams also evaluated the plane.

Unfortunately, the first prototype was lost in a reliability test accident in September 1967. Fortunately, both pilots

CL-84 completing its vertical ascent stage. (Canadair Photo)

ejected safely from the plane. The aircraft was flying at 3,200 feet at 150 knots in a forward velocity mode when the plane yawed to the left and quickly pitched downward. The investigation that followed identified the probable cause as a propeller failure. The plane was on its 306th test flight when the incident occurred. The program, though, went on with the construction of the three additional versions which incorporated a number of design changes from the original. Only two of them would actually fly. Two of the planes would be involved in non-fatal accidents due to mechanical problems.

The additional prototypes were built between February 1968 and February 1970. The testing of them would continue until 1974, over 20 years since the design work had started.

Two years before the end of the program, the CL-84 was demonstrated to the US Navy. Most significant was a flying demonstration that took place off a hundred-square-foot pad at the Pentagon. The *USS Guam* also hosted a number of both CL-84 STOL and VTOL flights from its deck.

Even though the model had achieved considerable success, neither the United States or the Canadian government showed enough interest to bring the program into production.

Even though two of the prototypes would have non-fatal crashes, the CL-84 was overall considered a success.

And today, one of those planes has survived, the second of the three prototypes produced, which is on permanent display at the National Aviation Museum of Canada in Ottawa. This particular plane made 196 flights with almost 170 flight hours. The plane was donated to the museum by Canadair in 1984.

4
Tilt-Thrust Systems

The Tilt Thrust concept covers a very large number of related VTOL concepts. They are all associated in that a single source provides power for both hover and cruise, and something rotates in order to point the thrust down, forward, or anywhere in between. Aircraft of this concept fall into four categories:

Tilt Engine - The only known example is the Bell Air Test Vehicle. This relatively simple aircraft had a jet engine mounted to each side of the fuselage. The engines pivoted between vertical and horizontal positions. The German VJ-101C also used the tilt engine concept. However, it also used lift engines, and the authors chose place it in another concept.

Tilt Duct - This group includes the Bell X-22, Doak 16, and Nord 500. These aircraft had propellers enclosed in ducts. The entire prop and duct assembly rotated. The duct greatly increased the prop efficiency and acted as lifting surfaces in forward flight. The engines were located in the fuselage or on the inboard portion of the wing.

Tilt Rotor - This group includes the Transcendental Model 1-G, Bell XV-3, Bell XV-15, Bell/Boeing V-22 Osprey, and the Bell 609 Civil Tilt Rotor. These aircraft perhaps have turned out to be the most successful of any of the Tilt Thrust con-

cepts. They all feature tilting rotors at the ends of the wings. The rotors are somewhere between the size of helicopter rotors and conventional propellers. In hovering flight they function as rotors, but in forward flight they function as propellers. The first two had engines mounted in the fuselage. The last three have the engines mounted in nacelles on the wing tips, and the entire engine/rotor assembly pivots as a single unit.

Tilt Prop - This group included the Curtiss-Wright X-19 and X-100. These aircraft had tilting propellers at the ends of stubby wings. The wings were small because the wide propeller blades produced lift in the vertical direction, called "radial lift," when the aircraft flew in forward flight. The engines were mounted in the fuselage. The authors considered this a separate category from Tilt Rotor because the propellers were closer to conventional propellers than to helicopter rotors and because of the use of radial lift.

The V-22 is the first Tilt Rotor aircraft to enter production, and soon will enter service with the U.S. Navy, Marines, and Air Force. Most of the problems inherent in this concept were worked out through extensive testing of the XV-3 and XV-15 research aircraft. The Bell 609 will be a civilian offshoot capitalizing on the extensive experience Bell had with the Tilt Rotor concept. These aircraft are proof that technical problems can be resolved and a concept taken to production.

The Air Test Vehicle (ATV) shows its strange configuration in flight. The fuselage-mounted jet engines look as though they were an afterthought in the design. (Bell Photo)

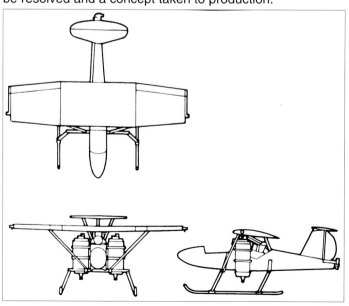

Three-view drawing of the Bell ATV. (Bell Photo)

The ATV on the ground, on the original skid landing gear, with engines in the horizontal position. (Bell Photo)

Later configuration featured a conventional landing gear. Note mission markings on the nose. (Bell Photo)

BELL AIR TEST VEHICLE TILT ENGINE VTOL (U.S.A.)
System: Built-from-scratch research vehicle using many existing assemblies
Manufacturer: Bell Aircraft Corporation
Sponsor: Company funded
Time Period: 1953 - 1955
Mission: Tilt Engine VTOL research
The Story:
The Bell Air Test Vehicle (ATV) was an experimental jet-powered VTOL that was a private development by Bell. Its purpose was to verify VTOL concepts that Bell engineers developed. Two Fairchild J44 jet engines, each rated at 1,000 pounds of thrust, provided horizontal and vertical thrust. They were mounted on the fuselage at the center of gravity and rotated between horizontal and vertical positions. It carried the civil registration N1105V. Overall dimensions were a length of 21 feet and a wing span of 26 feet. Gross take-off weight was about 2,000 pounds. The aircraft also was referred to as the Bell VTOL and the Bell Model 65.

For a VTOL aircraft, the ATV was a simple machine, to say the least. To save development funds, the open cockpit fuselage came from a Schweizer glider, the externally-braced wing from a Cessna 170, and the landing skid from a Bell Model 47 helicopter. Bell even borrowed the engines from the U.S. Air Force.

Conventional control surfaces were used during wing-borne flight. All controls were straight mechanical, with no stability augmentation or power boost of any type. A Turboméca Palouste turbine engine mounted above the fuselage just behind the cockpit provided 2.5 pounds of compressed air per second for pitch, roll, and yaw control in hover and low speed flight. A duct ran along the right side of the fuselage to carry the compressed air to the tail for pitch and yaw control. An internal duct ran the compressed air to each wing tip for roll control.

The ATV was constructed in eight months during 1953 and was completed in December. Tethered flights began the following January to assess hover characteristics. Within a month, a compressor failure in the right engine caused a fire and extensive damage, but the pilot was not injured and the ATV was rebuilt. The first free vertical flight was made on November 16, 1954. At some point during its flight test career, the helicopter skid was removed and the ATV was given a conventional landing gear and a tail wheel.

The flight test program ended in the spring of 1955. The ATV flew only 4.5 hours in about 30 flights during its career. No transitions from hover to conventional flight ever were performed. Following completion of flight testing, Bell donated the Air Test Vehicle to the Smithsonian Institution.

TRANSCENDENTAL MODEL 1-G TILT ROTOR VTOL (U.S.A.)
System: Built-from-scratch prototype
Manufacturer: Transcendental Aircraft Corporation
Sponsor: Private, with U.S. Air Force Support
Time Period: 1953 - 1956
Mission: Prove the feasibility and practicality of the Tilt Rotor concept
The Story:
The Transcendental Aircraft Corporation of Glen Riddle, Pennsylvania, was the first company to claim flying a successful Tilt Rotor aircraft. The Model 1-G was a small, high-wing experimental aircraft with a fuselage-mounted engine and fixed tricycle landing gear. The single pilot sat forward of the wing in a semi-enclosed fuselage. The clear plastic nose gave the pilot helicopter-like visibility. A single Lycoming O-290-A six cylinder engine produced 160 horsepower. The fuselage measured 26 feet long, and the aluminum wing measured 21 feet. The ailerons were fabric covered. Empty weight was 1,450 pounds and take-off weight was 1,750 pounds. The

projected maximum speed in helicopter mode was 120 miles per hour, and 160 miles per hour in airplane mode.

A 17-foot-diameter, three-bladed, fully articulated rotor was mounted at each wing tip. The rotor shafts tilted from pointing vertically for hover down to 6 degrees up from horizontal for forward flight. Electric motors controlled the tilt of the rotor shafts. Interconnecting shafts ensured that both rotors maintained the same tilt angle. Each rotor was driven through a two-speed gearbox. This allowed the pilot to lower the rotor rotation speed for more efficient cruise in forward flight.

Although the Model 1-G was a private development, the Wright Air Development Center at Wright-Patterson AFB issued contracts to study many of the Tilt Rotor's unique peculiarities.

The first was awarded in 1952 to investigate the dynamics and structural characteristics of the rotor system. The Air Force awarded a second contract in 1953 to investigate mechanical instability problems associated with tilting the rotors.

The first hover flight was either on June 15 or July 6, 1954 (references vary). The first forward flight in hover mode occurred on December 13, 1954, and the first forward flight with rotors tilted occurred only four days later. By April 1955, conversions with the rotors tilted up to 35 degrees from vertical were completed. Eventually, the Model 1-G completed numerous transitions up to 70 degrees of tilt with the wings sustaining over 90 percent of the weight.

The Model 1-G was destroyed during a test flight on July 20, 1955. After performing a virtually complete conversion, the friction lock on the collective pitch controller slipped, throwing the aircraft into an abrupt, steep dive. The pilot initiated a recovery, but there was not enough altitude to complete the pull-up before contacting the Delaware River. The aircraft flipped on its back, resulting in irreparable damage.

During the Model 1-G's brief career, it made over 100 flights and flew 60 hours. It demonstrated excellent controllability without vibration, and reached an altitude of 3,500 feet and an airspeed of 115 miles per hour in helicopter mode.

A second aircraft, called the Model 2, was to fly in late 1956. Transcendental received an Air Force contract in March

1956 and completed the Model 2 in October. Compared to the Model 1-G, it was stronger and more aerodynamic, but had the same basic configuration. The enclosed cockpit had side-by-side seating. Power was by one 250 horsepower Lycoming O-435-23 six cylinder engine. The wing was a foot longer, but the fuselage was four feet shorter. Empty and gross weights were 1,570 pounds and 2,249 pounds. Development ended when funding from Wright Air Development Center stopped. It could not be determined if the aircraft ever flew, and eventually it was dismantled.

DOAK 16 (VZ-4) DUCTED FAN VTOL (U.S.A.)
System: Built-from-scratch research aircraft
Manufacturer: Doak Aircraft Company
Sponsor: U.S. Army
Time Period: 1956 - 1972
Mission: Tilt Duct research
The Story:
Doak Model 16 was the first VTOL aircraft to demonstrate the tilt duct concept. It was built by the Doak Aircraft Company of Torrance, CA. The company president, Edmond R. Doak, had experimented with ducted fan and various other air moving principles since 1935. He first proposed a VTOL aircraft using the tilt duct principle to the military as early as 1950. The U.S. Army Transportation Research and Engineering Command purchased a single Doak 16, designated the VZ-4DA, serial number 56-9642. The Army issued the contract to Doak on April 10, 1956.

The basic configuration was a two-place tandem cockpit with a mid-wing, conventional tail, and fixed tricycle landing gear. The most notable feature of the aircraft was the ducted fan located at the tip of each wing. Conventional construction techniques were used throughout the aircraft. The fuselage was made of welded steel tubing and covered with a molded fiberglass skin from the cockpit forward. The aft fuselage, with its much straighter lines, was covered with thin aluminum sheet. The cantilever wing and tail unit were of all metal construction. Major design emphasis was on the wing tip ducts, fans, hovering stability parameters, power transmission system, and pilot controls. To save development costs, Doak in-

The little-known US Army Doak 16 used a Tilt Ducted Fan technique for its VTOL capabilities. (Doak Aviation Photo)

The Doak 16 in a captive hover test. (Doak Aviation Photo)

A shot of the Doak 16 during hover flight. Note the markings on the inside of the duck to measure tilt angle. (Doak Aviation Photo)

corporated numerous off the shelf items in the design, such as using the landing gear from a Cessna 182, seats from a F-51, duct actuators from T-33 electric flap motors, and the rudder mechanism from an earlier Doak aircraft.

Overall dimensions were 32 feet length, 25 foot, 6 inch wing span (including the ducts), 10 feet height, and wing area of 94 square feet. The design empty weight was 2,000 pounds with a design gross vertical take-off weight of 2,600 pounds. These grew to 2,300 and 3,200 pounds during the life of the program. To keep weight down, the original specification called for the aircraft fuselage to remain uncovered. However, it subsequently was felt that this would severely limit being able to obtain any meaningful forward speed data, and that the added weight would allow for much more valuable data to be collected.

The maximum speed was estimated at 200 knots, with a rate of climb at sea level of 6,000 feet per minute, 6,000 foot hover ceiling, 1 hour endurance, and 200 nautical mile range.

The wingtip-mounted ducts were five feet in diameter with a four foot inside diameter. Their construction was of aluminum alloy with a fiberglass leading edge section. Eight fixed pitch fiberglass fan blades turned at a maximum fan speed of 4,800 rpm. Ahead of the fan in the forward part of the duct were fourteen fiberglass variable inlet guide vanes. The vane angle varied during hover to modulate the thrust produced by the duct, and thus to obtain roll control. The prop was set back two feet from the front of the duct to prevent airflow separation. Nine stainless steel stator blades located aft of the fan straightened the air flow as it exited the duct.

The ducts rotated through 92 degrees, pointing horizontal for forward flight, and pivoting to 2 degrees aft of vertical during hover. The ducts rotated past vertical to compensate for the thrust from the jet exhaust. A switch on the control column initiated the duct rotation. To power the fans, drive

Doak 16 in free hover. (Doak Aviation Photo)

An early shot of a Doak 16 before application of fuselage skin. (Doak Aviation Photo)

shafts traveled through the wing quarter chord. Doak-designed flexible couplings compensated for misalignment and wing flexing.

A Lycoming T-53-L-1 turboshaft engine located in the fuselage just below the wing root provided power. It produced 825 hp (some sources stated 840 hp). A "T" box on the engine transmitted power to the ducts using a four-inch tubular aluminum shaft and two smaller steel shafts of 1.5 inch each.

Flight controls consisted of standard stick and rudder. An electrical and mechanical interlock system controlled all functions for both hovering and forward flight. There were no other cockpit controls. In hover, a cruciform shaped vane in the tail pipe at the rear of the fuselage controlled pitch and yaw by deflecting the engine exhaust. Rotating the inlet guide vanes in the ducts provided roll control by restricting airflow. As the ducts rotated from vertical to horizontal, a mechanical control system gradually phased out control of the inlet guide vanes and left them aligned with the duct airflow. There was no artificial damping or power boost. Doak looked down on any type of automatic stabilization system, feeling that the aircraft should be a satisfactory flying machine without any such equipment. Careful selection of the duct location allowed the fuselage to remain level throughout the transition.

Ground testing began at Torrance Municipal Airport during February 1958. Tests consisted of 32 hours in a test stand, and 18 hours of tethered hovering and taxi tests. The first free hovering flight was performed on February 25, 1958. Initial Doak testing at Torrance was completed in June 1958 and was followed by a complete tear down inspection. The aircraft then was transferred to Edwards AFB in October 1958. At Edwards, it performed 50 hours of tests, including transitions at altitudes as great as 6,000 feet. Following these tests, the Army accepted the Doak 16 in September 1959 and transferred it to NASA Langley for further tests.

The Doak 16 demonstrated conventional, vertical, and short take-offs and landings. While the aircraft exhibited some undesirable flight characteristics, only a few were considered fundamental to the Tilt-Duct system and these were solvable. One of the most undesirable characteristics was a nose-up tendency during transition from hovering to forward flight caused by the ducts. Short take-off and landing performance proved to be below expectation. At moderate speeds with partial duct angles, the load distribution across the wing was less than expected. This was because the ducts at the wing tips provided a large part of the total lift, especially at lower airspeeds. Greater drag results when more lift is produced outboard on the wings. The maximum speed demonstrated at sea level was 230 miles per hour, maximum rate of climb at sea level was 4,000 feet per minute, and the range was 230 miles.

By late 1960, Doak must have been in serious financial trouble. They sold the patent rights and all engineering files to Douglas Aircraft, in nearby Long Beach (Edmond Doak had worked at Douglas in the 1930s and certainly still had many friends there). Doak finally closed its doors early in 1961. Douglas liked the aircraft and had some ideas for improving it, primarily by installing a larger engine and making numerous structural improvements. They made an unsolicited proposal to the Army in 1961, but could not sell their ideas. The Doak 16 remained at NASA Langley until August of 1972. Eventually it was transferred to the U.S. Army Transportation Command Museum at Fort Eustis, VA, near Newport News, where it is on display.

NORD 500 DUCTED FAN VTOL (FRANCE)
System: Built-from-scratch research aircraft
Manufacturer: Nord Aviation
Sponsor: Company funded
Time Period: 1966 - 1971
Mission: Tilt Duct VTOL research
The Story:
The Nord 500 was a single seat, company funded research aircraft. Its mission was to evaluate principles of the Tilt Duct

Nord 500 in captive hover. (Nord Photo)

The unique look of the X-22 with its four ducted fan powerplants. (USAF Photo)

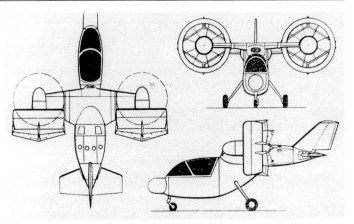

Three-view drawing of Nord 500 (Nord Photo)

propulsion concept for VTOL aircraft. The enclosed cabin contained an ejection seat. Two 317 horse power Allison T63-A-5A (or Allison T63-A5T, or 250-C18, depending on the source) turboshaft engines were located side by side in the rear part of the fuselage. They drove two five-foot diameter props through interconnected shafts. Moveable vanes in the propeller slipstream controlled the duct positions aerodynamically. There were no other mechanical controls for rotating the ducts. The ducts tilted, along with a short section of wing. During hover, control in roll was by differential thrust, while control in pitch was by collective tilting of the ducts. There was no provision for attitude control of the fuselage because the ducts pivoted freely. The intended top speed was 218 miles per hour.

The first prototype was completed in Spring 1967 and was used for mechanical and ground tests. The second prototype made its first tethered flight during July 1968.

Nord merged with the Aérospatiale Corporation in about 1970, and the aircraft became known as the Aérospatiale N 500. Although a more sophisticated and more powerful version was in planning, all efforts on the Nord 500 appear to have stopped by 1971.

BELL X-22 TILT DUCT VTOL (U.S.A.)
System: Built-from-scratch VTOL research aircraft
Manufacturer: Bell Helicopter
Sponsor: U.S. Navy
Time Period: 1962 - 1984
Mission: Tilt Duct technology demonstration and development
The Story:
The U.S. Navy's X-22 was developed as a V/STOL research aircraft. As such, it proved to be one of the most versatile and longest-lived of the many V/STOL aircraft that were developed. It also had the distinction of being the only aircraft to have a variable stability system incorporated into the basic design from the beginning. This feature contributed to its versatility and long service life by allowing it to perform V/STOL research that applied to a wide range of aircraft, not just to the peculiarities of the X-22 configuration itself.

The X-22's history goes back to the Tri-Service V/STOL Transport Program that addressed needs of the Army, Navy, and Air Force. One goal of this program was to develop a small number of prototype V/STOL transport aircraft that used different concepts and to perform operational evaluations of their usefulness. While the Air Force was interested in the tilt-wing concept, such as the XC-142, the Navy faced tougher problems caused by shipboard compatibility requirements. Studies showed that a duel tandem ducted fan configuration permitted a shorter wing span for a given weight, allowing a stubbier design that could fit on existing carrier elevators and would eliminate the need for complex wing folding mechanisms. The duct around each of the four props also would improve propeller efficiency and provide a safety benefit to personnel working in the cramped environment of a ship's flight deck.

The Navy awarded a $27.5 million contract for the design and development of two identical X-22s to Bell Helicopter of Niagara Falls, NY, in November 1962. Bell's internal designation was the Model D2127. Bell was no newcomer to the V/STOL business. They already were building the world's first commercially licensed helicopter. In addition, Bell built and flew the Air Test Vehicle and X-14 VTOL research aircraft, and was at the mock-up stage for the XF-109, a supersonic V/STOL fighter that never was built. While the X-22 was to be a research aircraft, it was representative of a possible small V/STOL transport. Thus, it could carry a 1,200 pound payload and represent a commercial aircraft that could carry up to six passengers. Its length and wingspan were each a little over 39 feet, and maximum gross weight was 16,700 pounds.

The basic configuration of the X-22 was four ducted fans that could rotate together between vertical and horizontal positions for the various flight modes. Four General Electric YT58-GE-8B/D turboshaft engines rated at 1,250 horsepower each were mounted in pairs at each wing root. They powered a common drive shaft that turned all four props. The "B" and "D" designations on the engines referred to two engine configurations that differed only by their fuel controllers. These engines had both controllers and switched between them automatically based on whether the X-22 was operating in hover mode or cruise mode. 465 gallons of useable fuel was carried in fuselage tanks.

The power transmission system consisted of a total of ten gearboxes. It reduced the engine's nominal 19,500 revolutions per minute speed down to the propellers' nominal 2,600 revolutions per minute. This arrangement also allowed all four props to continue operating with any number of engines failed or intentionally shut down.

Hamilton Standard built the four propellers. The 7-foot diameter, 3-bladed props were fabricated of fiberglass bonded to a steel core, making them 25 percent lighter than metal props yet giving them three times the fatigue strength. A nickel sheath was mounted over the leading edge. Very high prop efficiency was achieved by placing the props inside of the ducts, so much so that the X-22 could still take off on three engines, fly on two, and make a conventional landing with only one. The two forward ducts were mounted to small pylons on the forward fuselage, and the two rear ducts were mounted to stubby, dihedralless wings on the aft fuselage.

Hydraulic actuators rotated each of the four ducts, but mechanical and electrical interconnections insured that all rotated together. The X-22 performed a vertical take-off with the ducts in the vertical position and then transitioned to wing-borne flight by rotating the ducts forward. The ducts acted as wings when rotated to the horizontal position for forward flight. In this mode, the X-22 basically had the efficiency of a canard design, rather new and radical for the 1960s (actually not, the Wright Brothers' first aircraft all were canard designs!).

All four props operated from a common drive shaft and thus always turned at the same speed. With each prop turning at high speed, very quick and precise thrust control could be obtained by varying the blade angle. Four elevons, one placed at the rear of each ducted fan assembly, were the only control surfaces. Placing the elevons in the prop slipstream made them very effective, even at low airspeeds. Despite the significant looking vertical stabilizer, there was no rudder. Movement of the elevons and changes to the prop pitch achieved all flight control.

Flight control in horizontal flight was achieved using a conventional looking control stick for pitch and roll. Moving the stick caused the elevons to move, either differentially between the front and rear for pitch, or differentially on left and right sides for roll. Yaw was achieved by moving the rudder pedals, which changed the propeller blade angles to produce differential thrust. There were also throttles for each engine and a lever to control the angle of the ducts. In forward flight, the front ducts were rotated to 3 degrees up from horizontal and the rear ducts rotated to 2 degrees below horizontal. This gave an optimum incidence of 5 degrees between the two pairs.

While hovering, the pilot used the same control stick to control pitch and bank motions. Stick inputs caused the flight control computer to command minute changes to the prop

blade angles to vary the thrust, causing the X-22 to tip forward, aft, or sideways. The thrust forces now being off vertical caused the aircraft to move in the appropriate direction. Yaw was controlled by moving the rudder pedals, but now the computer caused differential movements of the elevons between the left and right sides. The pilot also could rotate the ducts to assist the fore/aft motion during hover.

During transition, with the ducts at some intermediate angle, the pilot's control inputs produced mixed propeller pitch and elevon deflections. The ratio of mixing between the props and elevons was a function of the duct angle. The ducts rotated at 5 degrees per second.

The X-22's flight controls also included a variable stability system. This was another flight control computer that modified the basic airplane responses so that the characteristics of other aircraft, either real or imagined, could be produced. In today's terminology, it would be called an in-flight simulator. Every airplane is unique, having its own set of flight characteristics. The variable stability system followed algorithms that were developed specially for each test and programmed into the computer. They produced extra control surface motions that caused the X-22's flight characteristics to be varied, thus producing motions that are not characteristic to the X-22 airframe, but rather to the aircraft being simulated. This gave the X-22 the capability to perform research that would be applicable to a broad range of other aircraft, not just the unique characteristics of the X-22 itself.

The Calspan Corporation of Buffalo, NY (then known as the Cornell Aeronautical Laboratory), designed the variable stability system. Calspan had a long history of developing and operating in-flight simulators for the Air Force. Some of the requirements for incorporating variable stability placed difficult design requirements on Bell. For example, there had to be virtually no hysteresis in any of the flight controls. To explain hysteresis, think of the pilot moving the control stick and then letting the stick go. The stick and surface may not

X-22 in forward flight. (Bell Photo)

return to the exact starting point after the pilot takes his hand off the stick. This is the result of friction or surface irregularities in the mechanical connections and hinges. Hysteresis is a measurement of how far a moving component ends up from its original starting point. For normal aircraft, an error of a few percent is tolerable and easily compensated for by the pilot. However, in order for the variable stability system to function properly, the hysteresis has to be virtually zero so that the computer knew the precise location of the controls. Bell engineers accounted for this and other Calspan requirements in their design, assuring the proper operation of the X-22 for variable stability programs.

When operating in the variable stability mode, the pilot in the left seat would experience different flight characteristics. The pilot in the right seat served as the safety pilot. The X-22 always had the same X-22 characteristics when flown from the right seat. Thus, the safety pilot always knew exactly what flight characteristics to expect when taking over control from the left-seat pilot.

The first X-22, tail number 1520, rolled out on May 25, 1965, and was followed by fifty hours of propulsion tests in a test stand. The first flight in hovering mode was not made until March 17, 1966. On this 10-minute flight, four vertical take-offs and landings and a 180 degree turn were made. It then performed a series of STOL take-off and landing tests with the ducts tilted at 30 degrees. Unfortunately, the first X-22 was damaged beyond repair on its fifteenth flight on August 8, 1966. Although it had flown only 3.2 hours, it suffered a dual hydraulic failure about four miles from its base at Niagara Falls Airport. The first transition from wing-borne flight to vertical flight was made under the high stress of an emergency landing. The fuselage broke in half, with the rear section coming to rest inverted. While the aircraft was lost, neither pilot was injured. This event gives an interesting example of the principle of dual redundancy, because the two systems were identical and suffered the same failure within minutes of each other. Swivel fittings were used in the ducts to provide hydraulic fluid to the elevon actuators. Both failed due to excess vibration. The fix included replacing the swivel fittings with loops of flexible tubing, replacing the aluminum hydraulic lines with ones made of stainless steel, and placing additional clamps on the hydraulic lines to minimize vibration.

The second X-22, tail number 1521, flew on January 26, 1967. With pilots from Bell, the Army, Navy, and Air Force, the X-22 flew frequently over the next several years. At the completion of the Tri-Service testing in January of 1971, the X-22s completed 228 flights, 125 flight hours, performed over 400 vertical take-offs and landings, over 200 short take-offs and landings, and made over 250 transitions. It also hovered at 8,000 feet of altitude and achieved forward speeds of 315 miles per hour. These flights demonstrated that the X-22 had good basic stability and that vertical take-offs and landings could be performed easily. Operation in ground effect was a little less stable, but still positive. Hovering was easier than in most helicopters. In horizontal flight, all responses to pilot

control inputs were excellent. Transitions were accomplished with minimum pilot workload. Landing position could be controlled precisely. The aircraft's stability augmentation system helped the pilot immensely during transition and hover. The aircraft was still controllable without augmentation, but required a significant increase in pilot workload. This system provided rate damping in pitch, roll, and yaw only during hover and low speed flight.

Numerous problems were discovered and fixed during these flights. These should not be taken as design faults, since the purpose of building a research aircraft is to test new concepts and see how well they work! While all were fixed, they were not necessarily fixed in an optimum manner, as would be done for a production aircraft. Further development may have provided an optimal solution. But, being a research aircraft, a fix that worked for the intended mission was good enough. Some of them included the following:

• Failure of the linkage synchronizing the front and rear duct angles, resulting in the front ducts rotating to 30 degrees while the rear ducts remained vertical. Fortunately, this happened on the ground. The aluminum shaft that transmitted the proper duct angle was replaced with one made of stainless steel.

• In forward flight with the ducts near the stall angle of attack, the airflow from the lower lip separated from the duct surface as it entered the duct, causing a very loud buzzing as the turbulent flow hit the prop. Installing a number of vortex generators on the bottom inner-lip of each duct reduced this problem significantly.

• A number of fatigue cracks developed on the inside of the duct skin and ribs. Apparently this was caused by a wake of very low pressure being pulled behind each prop blade. Thus, for each revolution of the prop, the surface was hit with three pulsations of high, then low pressure. This was corrected by replacing the ribs with slimmer ones, then building up an eighteen-inch wide ring of fiberglass inside each duct,

With ducts at about a 45 degree angle, this X-22 is transitioning to horizontal flight. (Bell Photo)

maintaining the 3/8-inch clearance between the prop tip and the wall.

With the Navy satisfied with the basic operation, they awarded a contract to Cornell in July 1970 to operate and perform flight research using the X-22, with particular emphasis on operating in the variable stability mode. Over the next ten years, Calspan flew five test programs as summarized below:

• August 1971-February 1972: evaluation of steep STOL approach paths of 6-10 degree glide slopes, at airspeeds of 65-80 knots, with a variety of oscillatory characteristics in the pitch mode.

• June 1972-February 1973: continuation of previous effort, but with variation of roll and yaw oscillatory characteristics.

• October 1973-April 1975: evaluation of control, display, and guidance requirements for STOL instrument approaches. Determined desirable control system requirements and information displays for pilot use to permit transition from forward flight to a decelerating steep approach and then to a hover at 100 feet, followed by a touchdown, all under instrument conditions.

• February 1977-March 1978: expanded the previous experiment by evaluating the usefulness of a head-up display for STOL instrument approaches. Used precision radar distance measuring equipment to establish the aircraft's position within 3 inches. Evaluated a variety of display formats that presented data to the pilot, such as attitude, airspeed, altitude, horizontal location, and range to touchdown. Also, simulated the AV-8B, a specific aircraft, rather than a generic set of aircraft characteristics.

• November 1978-May 1980: generated flying quality and flight control design requirement data for V/STOL aircraft performing shipboard landings. In performing this task, typical shipboard pitching and rocking motions had to be added to the guidance beam, then various compensation schemes were programmed into the flight computer on the X-22. Hover, and then simulated touchdowns at 100 feet were performed.

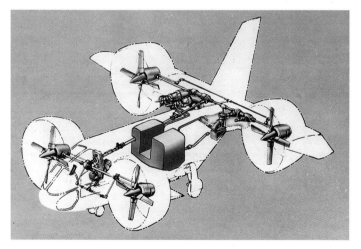

An X-22 cutaway view showing the numerous gear boxes needed to transmit power from the engines to the props. (Bell Photo)

By the time the last research program was completed, the military's interest in V/STOL research was all but gone, and the program office within the Navy that oversaw the X-22's testing was disbanded. Calspan sought added research programs, but the most they could accomplish was to get the Naval Test Pilot School to use the aircraft for some V/STOL demonstration flights for their students during 1981 and 1982. The aircraft made its last flight in October 1984. Ownership was transferred to the Naval Aviation Museum at Pensacola, Florida, but the museum never had any desire to display unique aircraft that were not typical of Naval aviation use. The X-22 remained in storage at Calspan's facility at Buffalo Airport in hopes that further projects would return this unique test vehicle to service, or at least be acquired by an aviation museum.

No further work ever arose, and many efforts to transfer the X-22 to an appropriate museum in the western New York area fell through. In 1995, Calspan moved it outdoors because they needed the hangar space. To protect it from the elements, the Buffalo & Erie County Historical Society paid to cover the X-22 in a plastic wrapping. In 1998, the newly formed Niagara Aerospace Museum at Niagara Falls, NY, acquired the X-22 and placed it on display.

BELL HELICOPTER XV-3 TILT ROTOR VTOL (U.S.A.)

System: Built-from-scratch research aircraft
Manufacturer: Bell Helicopter
Sponsor: U.S. Air Force and U.S. Army
Time Period: 1951 - 1966
Mission: Tilt Rotor research aircraft
The Story:
By the late 1940s the unique capabilities of helicopters became recognized and were being exploited for both military and civilian purposes. As their use expanded, limitations also rapidly became apparent. Most noted was their limited top speed, resulting from their inability to counteract the roll at higher speed that resulted from the rotor advancing into the airflow on one side, while retreating on the other. One attempt to counter this phenomenon was the compound helicopter that used small wings to pick up the flight loads as forward speed increased. This reduced the loading on the rotor, and thus the rolling tendency at higher speeds. An alternate solution was the Tilt Rotor. This concept used a more or less typical helicopter rotor to take off and land, then tilted the rotor forward to serve as a propeller for conventional forward flight. The Tilt Rotor, with its rotor facing straight into the airflow, had the potential to allow a much higher forward speed than would the compound helicopter.

Bell had worked on the Tilt Rotor concept since the late 1940s. They came to believe that the idea not only was practicable, but also was well within the existing state-of-the-art. In August 1950, the Air Force and Army announced an official Tilt Rotor design competition. Bell's proposal for their Model 200 won the competition for the XV-3 (originally designated the XH-33), and Bell was awarded a contract in May

1951. The XV-3 was built and initially flight tested at Bell's Ft. Worth, Texas, facility. Two prototypes were built, carrying tail numbers 4147 and 4148.

The basic configuration was a metal fuselage with the slender metal wing mounted mid fuselage. A large helicopter-like rotor was mounted on each wing tip. The rotor shafts were oriented in a vertical position for take-off, landing and hover like a helicopter, and moved to the horizontal position for forward flight like a conventional aircraft. A single Pratt & Whitney R-985-AN-1 Wasp Junior piston engine, producing 400 horsepower for flight and 450 horsepower for take-off, was located in the fuselage just behind the wing. Although a more powerful engine would have been desirable, the Wasp Junior was picked because of its proven reliability. A series of gearboxes and drive shafts transferred power to the rotors. Fuel was stored in a 100 gallon tank located in the fuselage just under the wing. The fuselage length was 30 feet, 4 inches, the wing span was 31 feet, 3 inches, the height was 13 feet, 7 inches, and the wing area was 116 square feet. The maximum speed was predicted to be 136 knots.

Three-bladed, fully articulated, 25-foot-diameter rotors were used initially. They were driven through a two speed transmission that could be shifted to a lower gear to allow the rotors to turn at a lower speed while the engine maintained a higher speed for greater cruise efficiency. The rotor tilt was controlled by an electric motor enclosed in a fairing at each wing tip. Each rotor could rotate through a 90-degree arc from vertical to horizontal in 10 to 15 seconds. The pilot could stop or reverse the rotor motion at any point. Steady, stable flight could be maintained with the rotors in any intermediate position.

Although the XV-3 was intended as a research vehicle to evaluate the Tilt Rotor concept and to provide design and test data, military contracting rules at the time required that it also be designed and demonstrated with a military mission capability. Thus, the XV-3's military mission was observation, reconnaissance, and medical evacuation. To accomplish this mission, it had mounting points for two litters. For military missions, the maximum range was calculated to be 480 nautical miles with a pilot, two passengers, and 67 pounds of cargo. For a 100 nautical mile combat radius, it could carry the pilot, three passengers, and 217 pounds of cargo.

During development, wind tunnel tests of a powered _ scale aerodynamic and dynamic model were run in the Wright Field 20-foot wind tunnel in June-November 1951 to determine stability and control characteristics, and performance capabilities in all modes of flight. Tests confirmed that good propulsive efficiencies for cruise could be obtained using helicopter rotors as propellers by reducing the rotational speed to about half of that used for hover. Only minor design changes were recommended as a result of these tests. What the wind tunnel tests did not look at was the impact of the large, slow turning rotors on the aeroelastic and longitudinal stability of the aircraft at cruise speeds. Aeroelastic effects, which are the coupling of the bending motions of various flexible com-

XV-3 Tilt-Engine XV-3 VTOL. (US Army Photo)

ponents on the aircraft, plagued the XV-3 throughout its early life. The analytical methods to predict such phenomena had not yet been perfected, and the need for such testing was not recognized until it was encountered in actual flight.

However, engineers knew about elastic modes, and rotor whirl mode tests were performed using the engine and drive system in hover mode in an outdoor test facility. These tests duplicated the aircraft drive system from the engine to the rotor as closely as possible from both a mechanical and structural standpoint. Some rotor mechanical instabilities were encountered and corrected, then a 100-hour endurance test was performed.

The first XV-3 (serial number 4147) was constructed between January 1952 and December 1954. The aircraft rolled out on February 10, 1955. Following roll out, airframe and control system proof load tests, vibration surveys, and system functional tests were performed. Then, with the XV-3 mounted on a tie down stand, full conversions and operation in helicopter and airplane modes were tested. Ground runs included a rotor stability survey and ten hours of full power operation. These ground tests were completed by August. No major difficulties were noted, and rotor stability checks showed the aircraft to be free of any resonance tendencies.

The first hover flight was made on August 11, 1955, by a Bell test pilot, but limited progress was made in flight testing during the next three years due to continuing problems with wing/pylon/rotor instabilities during flight. The first of these occurred after only one week and 1.2 hours of flight time, during an air taxi test. It resulted in a hard landing, and the XV-3 sustained rotor and airframe damage. While the damage was not extensive, the discovery of the instability raised major concerns. Repairs were made, and several modifications also were made which were felt would eliminate the elastic coupling problem. These included stiffening the rotor controls and adding external struts to make the wing more rigid. The trailing 25 percent of the wing was made to droop during hover to reduce disturbance to the rotor airflow during hover, and also to reduce stall speed to 85 knots, making transitions at a lower speed possible. 200 hours of ground runs and tie down tests were made to evaluate these modifications before clearing the XV-3 to fly again. The next hover flight finally occurred on March 24, 1956.

Confident with the ground test and hover results, Bell finally began envelope expansion flights in June 1956 and performed a nacelle tilt of 5 degrees on July 11. By July 25, they reached 70 degrees tilt and 80 knots forward speed when another rotor instability was encountered. Flight testing resumed on September 26 after more modifications were made to the rotor system and ground run evaluations were completed. On October 25, during another test flight, 4147 encountered another very severe inflight rotor instability and crashed. The pilot was seriously injured, but survived. Bell took a serious look at the entire rotor system, and decided that the basic design and characteristics of the three-bladed articulating rotor system were unsatisfactory for the XV-3.

Prior to resuming flight testing with the second prototype, 4148, Bell made numerous changes. They kept the R-985 engine, despite its limited power, because of its very good reliability record. The flush engine cooling air inlet was replaced with a scoop that spanned across the top of the fuselage, from one wing root to the other. As had been done with 4147, the rotor controls were stiffened and the wing was braced with external struts on the bottom. The #2 XV-3 was shipped to NASA's Ames Research Center, where it flew in their 40 x 80 foot wind tunnel in September and October 1957. Serious flutter problems with the original three blade rotors were confirmed, and Bell decided to replace them with two-bladed, semi-rigid type rotors of 24 foot diameter, mounted on shorter masts. The semi-rigid, versus the fully articulated design, in itself further increased rigidity, which would decrease the possibility of aeroelastic coupling. While in the wind tunnel, pilots were able to practice conversion procedures and gear changes. Further changes to the two bladed rotor/pylon design were still required to eliminate pylon oscillations, but these changes were tested while still in the tunnel, and again appeared to eliminate the problem in all conditions that were tested. The pilots found conversion to be quite easy, but engine gearbox shifting proved to be difficult, requiring considerable manipulation of the pitch controller and throttle throughout the twenty second process.

Flight testing of 4148 began at Bell on January 21, 1958. Conversions to 30 degree pylon angle and speeds up to 110 knots were accomplished by April 1, 1958. Pilots also demonstrated autorotation to a landing following a simulated engine failure. Helicopter characteristics were rated as good, especially in high speed forward flight, and vibration levels were lower than expected. By May 6, another rotor oscillation was encountered in flight at a 40 degree pylon angle, and the aircraft again was grounded. Bell, the Army, and NACA decided that another series of wind tunnel tests was in order. Unfortunately, the 40 x 80 foot tunnel at Ames Research Center was heavily scheduled through the summer, and the XV-3 sat grounded until being shipped to Ames in September. During this time, Bell conducted analog computer simulations to further analyze the instability problem and recommend configuration changes prior to beginning the wind tunnel tests.

During October 1958, the XV-3 finally went into the 40 x 80 foot wind tunnel at NASA Ames, and more refinements resulted. Changes included increasing the control system stiffness to three times greater than original, adding counterweights in to the rotor collective control mechanism, and increasing the blade sweep angle.

Flying resumed on December 11. On December 17, the XV-3 achieved 30 degrees tilt, and on December 18 achieved 70 degrees tilt. After making a few minor rigging corrections, a second flight was made and a full conversion to 90 degrees finally was achieved. The conversion was made in steps, starting at 90 knots in full helicopter mode and finishing at 115 knots in full airplane mode. A second full conversion was made the next day. This was three years after the date projected for full transition at the time of the XV-3's roll out. Over the next ten weeks, final aerodynamic and control system refinements were made, including adding a large plywood ventral fin to improve the poor directional stability.

With all the changes that were made to the rotors and pylons, Bell tried something new to see if the wing really was contributing to the instability. The strut attachments were modified with locking devices that allowed the XV-3 to fly with the supporting struts either locked in place to maintain wing rigidity or loose to provide no support to the wing. They were hydraulically actuated, and the pilot could switch between the two settings in flight. Bell assumed that if the instability resumed with the struts unlocked, then the wing indeed did contribute to the problem. The new device was tested on January 16, 1959, at rotor angles up to 85 degrees from vertical with no instability problems. On the January 22, the struts were removed and the XV-3 flew up to 120 knots with the rotors 50 degrees from vertical, again with no instability noted.

A wheeled landing gear was added to the landing skids, and short take-offs were made on April 13, 1959, using less than 200 feet of runway, and using only two thirds power. Optimum rotor tilt angle was found to be ten degrees forward of vertical.

The XV-3 was characterized with its long protruding greenhouse canopy. (Bell Photo)

The XV-3 had the look of both a helicopter and standard horizontal take-off aircraft. (Bell Photo)

The first in-flight gear change was made on April 14, 1959, and a forward speed of 120 knots was achieved at a much lower engine speed with an accompanying lower vibration level. The gear change took about ten seconds to accomplish. The process was very similar to shifting a manual transmission in a car, requiring manipulation of the aircraft's collective control, throttle, and clutch.

By April 24, 1959, the XV-3 finally was ready for formal evaluation by the military. It was shipped to Edwards AFB for a two month flight evaluation that began on May 14, 1959. The joint Air Force and Army evaluation consisted of 38 flights and a total of 29.6 hours. Forty conversions were made, as were twenty gear shifts to lower the rotor speed while in airplane mode. Also demonstrated were power-off conversions from airplane mode to helicopter mode followed by autorotation to a safe landing. Flights up to 12,000 feet were performed.

Air Force and Army evaluators concluded that conversions could be performed easily at all airspeed and fuselage attitudes that were tested. They considered the concept to be operationally practical because of low down wash velocity and temperature, low vibration, reasonable noise levels, and excellent reliability. The XV-3 demonstrated good behavior during stalls, good rolling take-off performance, and good basic controllability without electronic or mechanical stability augmentation.

On the negative side, items related to the prop rotor concept included an erratic lateral darting tendency and roll oscillations during hovering in ground effect. A large increase in power was needed as hovering flight was approached. Weak longitudinal and lateral-directional stability was also observed at low speed in helicopter mode, as was excessive blade flapping during longitudinal and directional maneuvering in airplane mode. There was high parasite drag in all configurations at high speeds. And last, the XV-3 displayed a fore/aft

surging motion, especially severe in rough air, attributed partly to the use of large, lightly loaded rotors as propellers.

Other shortfalls were typical of an aircraft designed as a research vehicle. These included performance being significantly less than predicted and weight growth. The design empty weight was 3,500 pounds and grew to 4,200 pounds, this before the Air Force added 160 pounds of instrumentation. At a gross weight of 4,800 pounds, only the test pilot and about 50 gallons of fuel (half the fuel tank capacity) could be carried. This gave a maximum endurance of one hour.

In airplane mode in high gear and full power at 4,000 feet of altitude, cruise speed was 102 knots true airspeed and wing stall was 100 knots. In low gear and full throttle, cruise increased to 115 knots and wing stall was reduced to 94 knots. The XV-3 was dived to 155 knots, which was the top speed due to limits of collective pitch. Short period longitudinal dynamic stability began to deteriorate at 120 knots, and was considered unacceptable at 130 knots.

At the completion of Air Force testing, the XV-3 was shipped to NASA Ames where it remained on flight status through July 1962. By the time its flying career ended, the XV-3 had been flown 270 times by 11 pilots for a total of 125 hours. 110 full conversions were made by nine different pilots, six of whom performed the conversion on their first flight.

Flight testing of the XV-3 clearly demonstrated that technology had not yet developed an understanding of the aeroelastic coupling that could occur between the wing and a large, slow-turning rotor located at the wing tips. It wasn't until the mid-1960s that engineers finally had a relatively complete understanding of the coupling problem, following the development of sophisticated analytical tools and more capable computers. Although the XV-3 demonstrated the feasibility of the Tilt Rotor concept, it was limited because of low power, an unsophisiticated flight control system, and relatively low twist, helicopter-like rotor blades. Several times Bell proposed replacing the powerplant with a Lycoming T-53 turboshaft engine of 600 horsepower and making rotor and control modifications to eliminate the concept-related deficien-

XV-3 hovering in front of flight hangar at NASA Ames. (NASA Photo)

XV-3 starting conversion to horizontal flight. (Bell Photo)

cies identified during the Air Force evaluation, but the XV-3 never flew again.

However, the XV-3's career was not over yet. Under a NASA contract, Bell continued studying the phenomena of aeroelastic coupling and ways to improve high speed flight. Two additional sets of wind tunnel tests using the XV-3 were performed. The first was done in July 1962, evaluating several changes aimed at increasing maximum speed, improving high-speed flight characteristics, and decreasing the rotor flapping that occurred during maneuvering. The tunnel was run at 130 knots before instabilities set in. Following modifications, the instabilities did not return until 160 knots.

Using these results, Bell felt that they finally had a good understanding of the problem. In May 1965, the XV-3 returned to Bell, where further study, modifications, and ground runs were performed between July 1965 and March 1966. It then returned to Ames and again entered the 40 x 80 foot tunnel for the fourth, and last, time in May 1966. This time, the XV-3 was tested to 197 knots, the limit of the wind tunnel, without encountering any of the oscillations that had plagued the aircraft throughout its career. However, on May 20, while running at maximum tunnel speed and taking the last planned data point, both rotors tore loose following a wingtip fatigue failure, damaging the aircraft and permanently ending the XV-3's career.

Following these events, the XV-3 was placed in storage for many years, spending time at Wright-Patterson AFB and Davis-Monthan AFB. Eventually the U.S. Army Aviation Museum at Fort Rucker, Alabama, acquired it for display.

Although the XV-3 itself never achieved the promised high forward speeds that had been hoped for, it did prove that the rotor could be tilted, and thus free the helicopter from the limited performance inherent in the rotor being in the horizontal plane. The XV-3 demonstrated the basic practicality and technical advantages of the Tilt Rotor concept for use on a VTOL transport aircraft. It provided a firm data base with invaluable information for the eventual XV-15 and V-22 programs.

BELL HELICOPTER TEXTRON XV-15 TILT ROTOR VTOL (U.S.A.)

System: Built-from-scratch research vehicle
Manufacturer: Bell Helicopter Textron
Sponsor: U.S. Army and NASA
Time Period: 1972 - present
Mission: Tilt Rotor research aircraft
The Story:

While the XV-3 program was still working out its problems in the early 1960s, Bell already was looking to the future, convinced that eventually they would prove the viability of the Tilt-Rotor concept. In 1965, the U.S. Army issued a Request for Proposal for what it called the Composite Aircraft Program. Composite, in this case, had nothing to do with materials, but was for a vehicle that would have both helicopter and airplane characteristics, specifically, looking for a single aircraft to replace both the CH-47 helicopter and the C-7 Caribou. Three contractors were selected to perform design studies in 1966. From this, Lockheed and Bell were chosen to perform further exploratory definition studies, which were completed in September 1967. Bell's tilt-rotor design was designated the Model 266, and Bell performed wind tunnel tests of a .133 scale, semi-span, aeroelastic model and a 1/12 scale, semi free-flight dynamic model to verify their calculations. However, the Army dropped the development due to limited funds.

Following termination of the Composite Aircraft Program, Bell decided in 1968 to continue development for a proposed civil tilt-rotor aircraft, designated the Model 300. Initial work led to the design for a 9,500 pound aircraft powered by two Pratt & Whitney PT-6 engines powering 25-foot diameter rotors. One-fifth scale aerodynamic and aeroelastic models were built and tested extensively from 1969 through 1973. Full size rotor and rotating mechanisms were whirl tested to determine their hover performance and then tested in 1970 at the NASA Ames 40 x 80 foot wind tunnel at various rotation speeds, angles, and airspeeds up to the maximum tunnel

XV-15 mid way through a transition (Bell Photo)

speed of 200 knots. The rotor met or exceeded all performance and stability predictions.

Then, in 1972, NASA and the U.S. Army Air Mobility Research and Development Laboratory jointly started the Tiltrotor Research Aircraft Program. Two phases were planned, a Proof-of-Concept phase and a Mission Suitability phase. The objectives of the Proof-of-Concept phase were to verify the rotor/pylon/wing dynamic stability, explore the limits of the operational flight envelope, establish safe operating limits, assess handling qualities, investigate gust sensitivity, and examine the effects of disc loading and tip speed on downwash, noise, and hover operation. The objective of the mission suitability phase was to assess the application of tilt rotor technology to satisfy military and civil transport needs. Emphasis was to be placed on eliminating XV-3 deficiencies, obtaining excellent hover performance (including single engine operation while out of ground effect), and developing a fail-operate flight control system that assured good handling qualities under all normal and failure conditions. All this was to be accomplished while avoiding the use of advanced technology in the design and manufacturing in order to minimize the cost and schedule risks that often followed the application of emerging technologies to new situations.

Since the Tiltrotor Research Aircraft Program was to be strictly a research program and would not lead to the production of an operational aircraft, costs were to be kept under control by not making weight minimization a major factor and encouraging the use of off the shelf components. Advanced technologies like fly-by-wire and composite structures were to be avoided. Weight growth and performance shortfalls would be tolerated in order to minimize cost and schedule impacts. Even the number of aircraft to be built was a factor. The two aircraft option was selected because of the high accident rate experienced by most other VTOL research programs.

Bell and Boeing Vertol each received contracts for three-month studies, and each submitted proposals. Boeing's losing proposal, designated the Model 212, utilized two modified Lycoming T53-L-13 turboshaft engines mounted on non-tilting nacelles at each wing tip. To save cost and development time, they would use the fuselage and empennage from a Mitsubishi Mu-2J executive transport.

Bell's proposal started with the Model 300's design and evolved it into Model 301. Bell kept the rotor and transmission, but replaced the engines with the more powerful Lycoming T-53 because of the requirement to hover with only one engine and the greater empty weight and useful loads required. Another benefit of the engine switch was that the T-53 already had an oil system that could operate with the engine pointed vertically, which had been developed for the CL-84 program. Bell's proposal was submitted on January 22, 1973, and comprised 300 volumes weighing 774 pounds. Bell's proposal was selected in April 1973. NASA awarded contract NAS2-7800 for $28 million for the final design, fabri-

cation, and preliminary testing of two XV-15s on July 31. The total estimated cost of the six-year program was $45 million.

The 42 feet long fuselage design basically was that of a conventional aircraft, the structure being of semi-monocoque, fail-safe construction, and fabricated using light alloy material. There was no fuselage pressurization, and the structure was stressed from +3 to -.5 G. The airframes were designed for minimum service lives of 1,000 flight hours over five years.

The tricycle landing gear came from the Canadair CL-84. It utilized Goodyear magnesium main and nose wheels, and Goodyear hydraulically operated magnesium/steel disc brakes. The full-swiveling nose wheel incorporated shimmy dampers and a centering device. It retracted into a bay forward of the cockpit. The main wheels retracted into external pods on each side of the fuselage. A switch on the main gear strut prevented inadvertent gear retraction and tilting of the pylons more than 30 degrees from vertical when the aircraft was on the ground. The landing gear was structurally designed to withstand a touchdown sink rate of 10 feet per second at full gross weight. A 3,000 pound per square inch nitrogen gas system provided for emergency extension in the event of a hydraulic failure.

The H-tail consists of a horizontal stabilizer with a vertical stabilizer on each tip. This configuration was selected to provide improved directional stability at and near zero yaw angles. Rockwell International's Tulsa Division built the fuselage and tail units under subcontract.

Two pilots sat side-by-side in Rockwell-Columbus LW-3B ejection seats. Visibility out of the cockpit was very good. The crew entered through a door on the right side of the cargo compartment. The flight deck was heated, ventilated, and air

The XV-15 during the vertical portion of its mission profile. The design was characterized by the large cones on the front of the engines. (Bell Photo)

Artist's concept of XV-15 in tactical situation. (Air Force Museum Photo)

conditioned, but not the cargo compartment. The cabin could accommodate nine personnel if not filled with test equipment.

The wing measures 32 feet across, has a constant chord measuring 5 feet, 3 in., and a resulting area of 169 square feet (one of the design requirements was that the XV-15 be able to fit in NASA Ames' 40 x 80 foot wind tunnel, which influenced the wingspan and rotor size). It is swept forward 6.5 degrees, not for any futuristic aerodynamic reasons, but to insure there would be adequate clearance when the rotor blades flex in airplane mode. Wing dihedral is 2 degrees. Along the trailing edge, a flap measuring 11 square feet occupies the inboard third, and a flaperon measuring 20 square feet occupies the outer two thirds. The large flaps can be deflected down to 75 degrees to help provide additional lift at low speeds. In hover, the flaps and flaperons deflect downward

to reduce slipstream interference by the wing. The problems with the wing/rotor/pylon stability that plagued the XV-3 were eliminated by designing a very stiff wing and nacelle/wing attachment, and by placing the rotor hub as close to the wing as possible.

Each wing holds two fuel bladders that form a single crashworthy fuel tank in each wing. Together they hold a total of 219 gallons. The pump in each wing tank is powered from a different electrical system. In the event of a pump failure, both engines can feed from the same tank, or in the case of an engine failure, one engine can feed from both tanks. Cross feeds activate automatically in the event of a pump failure to assure uninterrupted fuel flow to both engines. In the event of a complete loss of electrical power to both pumps, the engine driven pumps still can maintain adequate fuel flow.

An Avco Lycoming LTC1K-4K engine, a specially modified version of the standard T53-L-13B engine, is mounted at each wing tip. They are rated at 1,250 shaft horsepower for continuous operation, 1,401 shaft horsepower for 30 minutes, 1,550 shaft horsepower for 10 minutes for take-off, and 1,802 shaft horsepower for two minutes for emergency power. Power is transmitted from the engines to the rotors using a coupling gearbox and transmission, which reduce the engine speed of approximately 20,000 revolutions per minute down to a rotor speed of about 565 revolutions per minute in hover. The three-bladed, semi-rigid rotors measure 25 feet in diameter and have a 14-inch chord. They were made of stainless steel and have a large amount of twist. (In July 1979, Bell received a contract from Ames for preliminary design of a composite rotor blade that would offer improved performance and increased life expectancy, compared to the existing metal blades. A set eventually was tested, but did not work well.) There are no flapping hinges, which means that the rotors

The technology of the XV-15 directly contributed to V-22 which would follow. (Bell Photo)

are rigidly confined to the plane of rotation. The rotors can flap forward or aft as much as 6 degrees. To assure power to both rotors in the event of an engine failure, a shaft that runs through the wing interconnects the two transmissions. As a result of the interconnect, both rotors turn when the first engine starts. In the event of a double engine failure, both rotors will autorotate at the same speed. The nacelle tilt can be varied from horizontal to 5 degrees aft of vertical. Interconnected double ballscrew actuators operate the tilt mechanism in each nacelle. This assures that both nacelles always will be at the same position. The interconnected drive shafts and redundant tilting mechanisms permit single engine operation and fail-operate tilt capability.

The cockpit has dual controls and resembles a helicopter cockpit, including a collective stick. The flight controls are designed to permit single pilot operation from either seat. In airplane mode, the control columns and rudder pedals work conventionally. In hover mode, the stick functions as a cyclic pitch controller. The mechanical mixing unit does everything needed to convert the controls from the helicopter mode to the fixed wing mode. Control authority between helicopter and airplane mode is phased in as a function of the nacelle tilt angle. This includes changing the rotors from cyclic pitch control in vertical flight to constant speed control for fixed wing flight. In airplane mode, the collective lever can still be used as a power lever. Moving the collective lever causes the throttles on the center console to move.

Two switches, mounted on the collective lever and operated by the pilot's thumb, control the nacelle tilt angle. One pivots the nacelles from end to end in about 12 seconds and allows them to be stopped at any position. The other switch moves the nacelles between pre-selected angles of 0, 60, 75, and 90 degrees (relative to horizontal). To rotate the nacelles, electrical valves activate hydraulic motors. In the event of a complete electrical system failure, the pilot can manually open the valves using T-handles in the cockpit. This will drive the nacelles to the helicopter position.

Sperry Rand built the original navigation/guidance system. A digital computer provides navigation and control information to the pilot using advanced mechanical and electronic displays. The Calspan Corporation of Buffalo NY designed the Stabilization Control Augmentation System to improve its flight characteristics. The XV-15 does not incorporate fly-by-wire. Ailerons, elevator, and rudder are hydraulically boosted with a triple hydraulic system. They remain active in all flight modes.

The XV-15's empty weight is 9,570 pounds with a vertical take off weight of 13,000 pounds. This allows 1,100 pounds for instrumentation, 400 pounds for pilots, and 1,400 pounds of fuel, while leaving a few left over for growth. Original estimated performance included a maximum level speed of 330 knots, service ceiling of 29,000 feet, and a range of 500 miles. (None of these goals ever were achieved, but they certainly did not detract from the XV-15 achieving its primary objective of proving the practicality of the tilt rotor concept!)

To minimize development costs, iron bird flight qualification tests were not performed on the complete rotor/transmission/engine/flight control system. Each component was developed and tested individually.

XV-15 #1, #N702NA, rolled out at Bell's Arlington, TX, Flight Research Center on October 22, 1976. Ground runs began in January 1977, and included 100 hours of system qualification tests on an elevated test stand in both the helicopter and airplane modes to demonstrate that the aircraft met final flight qualification requirements. The first hovering flight was performed on May 3, 1977, followed by hover and low speed evaluations. This short test effort consisted of only three hours of hovering during May. No problems that warranted corrections were uncovered. Following these flight tests, the transmissions and rotors were torn down, inspected and reassembled.

Despite the successful early test flights, the XV-15 program did not fly again for almost two years. This was because of NASA's insistence on full-scale wind tunnel tests before attempting a conversion, and the reality that limited program funds precluded performing both wind tunnel testing and flight testing. The need for these full-scale wind tunnel tests was a major program issue. Bell's position was that these tests were appropriate for investigating some potential problems, but that such testing would not guarantee discovering all potential problems. Bell was confident that problems with the rotor/wing/pylon stability, which plagued the XV-3 throughout its career, had been eliminated in the XV-15 design, at least up to the 200 knot speed limit of NASA's 40 X 80 foot wind tunnel. Bell also felt that the XV-15's size, while only a few feet larger than the XV-3, put the rotor tips closer to the tunnel walls, making the test results less representative of the aircraft's true characteristics. Last, Bell felt there was potential to do structural damage. Being rigidly restrained in the tunnel, excessive forces unknowingly could be generated in the structure. Bell eventually lost the argument, and aircraft #1 was shipped by C-5A to NASA's Ames Research Center in March 1978 for wind tunnel tests. These tests were

The XV-15 in level flight. (Air Force Museum Photo)

conducted in the Ames 40 X 80 foot wind tunnel in May and June 1978. Twenty hours of tunnel tests were performed at airspeeds between 60 and 180 knots. Configurations consisted of the rotors in helicopter and airplane positions, and numerous intermediate positions that would be encountered during transition. No unusual characteristics were noted in any of the tests conducted.

Following the wind tunnel tests, the #1 aircraft was torn down and refurbished at NASA Ames. The second XV-15, #N703NA, was nearing completion. Since the program lacked funds to keep two aircraft on flight status, testing resumed with the #2 aircraft, beginning ground tests in August 1978 at Arlington. Numerous minor problems plagued the aircraft during these tests, including a stress corrosion crack in the left engine gearbox, a clutch misengagement, and foreign object damage within the transmission. It finally made its first hovering flight on April 23, 1979. Conversion tests soon began, starting by rotating the nacelles only 5° forward on May 5. Successive tests gradually rotated the nacelles closer and closer to horizontal, until the first complete conversion was made on July 24, 1979. The XV-15 also achieved a forward speed of 160 knots on this 40 minute flight. The gradual buildup testing verified that steady state flight was possible at any point during the conversion.

The Navy soon became interested in the XV-15. Because of continuous funding shortfalls, the Naval Air Systems Command began providing funding in 1979 and 1980 to insure the XV-15 flight testing would proceed up through the completion of envelope expansion flights. In exchange, the Navy would be allowed to perform flight evaluations.

Envelope expansion flights using the #2 aircraft to demonstrate higher speeds and system performance continued to be performed by Bell at their Arlington facility. System design criteria dictated that any single failure would not prevent the completion of a normal flight operation, and that any double failure would still permit the crew to eject (The rotor blades, rotor hub components, and transmissions were exceptions to this requirement. To verify that the probability of failure for these components was negligible, they were designed to much more conservative standards and tested extensively.). During the contractor test program, all potential failures were simulated in actual flight or on the ground. On December 5, 1979, an actual engine failure occurred when the turbine seized. The transmission interconnect system worked properly, and both rotors continued to turn as designed. The predicted speed of 300 knots true airspeed was demonstrated with maximum rated power at a 16,000-foot density altitude in June 1980.

The contractor flight test phase was completed in August 1980. The basic conversion corridor and airspeed/altitude envelope up to 16,000 feet was demonstrated. About 100 full conversions were made. Some resonance problems were uncovered, as is normal in any helicopter development, but they were nothing compared to the problems encountered on the XV-3, and were fixed quickly. The XV-15 proved to

Both XV-15s at NASA's Dryden Research Center in 1981. (Bell Photo)

have very good handling qualities. Aeroelastic stability in helicopter mode also was as predicted. Conversion proved to be very straightforward. Cockpit vibration and noise were very low, as was exterior noise. Although the horizontal gust response in airplane mode was unusual, it was considered acceptable, as was the overall ride quality. Upon completion of the contractor flights, XV-15 #2 was shipped to NASA's Dryden Flight Research Center for continued testing, where it was joined by aircraft #1. Both XV-15s then operated at Dryden for a short period.

XV-15 #1 returned to Bell in September 1981. Flight testing by both NASA and Bell continued into the 1980s, and the two XV-15s proved to be virtually free of any significant problems. Additional accomplishments that were demonstrated included:

• 1.7 hour cruise endurance in airplane mode.
• Cruise speed of 230 knots.
• Take off as helicopter, fly twice as fast as a helicopter, and deliver payloads on half the amount of fuel when traveling distances of greater than 115 miles.
• Autorotation descents in helicopter mode, but never to a full touchdown.
• STOL take-offs with the nacelles tilted between 60 and 70 degrees, at the maximum gross weight of 15,000 pounds.

For taxiing on wheels, it was found that tilting the nacelles forward of vertical only 1 degree was enough to start the XV-15 moving forward. Tilting the nacelles aft of vertical brings the aircraft to a quick stop. The XV-15 tends to rock a bit more than other aircraft because of the weight of the engines and props all the way out at the wing tips. The brakes are not powerful enough to allow instant stops, but powerful enough to use differentially to turn the aircraft.

In hover, roll control is provided by differential rotor collective pitch, pitch control by cyclic pitch, and yaw by differential cyclic pitch. For maneuvering in the hover mode, many of the maneuvers normally performed by moving the cyclic control are done by tilting the nacelles. A combination of rotor angle and cyclic pitch also is used to vary the pitch attitude without moving forward. By tilting the rotors forward and simultaneously putting in aft cyclic control, the nose will pitch down, giving improved visibility over the nose.

Vertical liftoff is very easy, even on a new pilot's first attempt. Whereas helicopters tend to lift off and promptly bank and pitch up or down slightly, the XV-15 holds attitude on liftoff. Lateral movement is accomplished by banking slightly so that the thrust now has a small side component. The XV-15 can translate sideways at 35 knots with no tendency to turn into the wind. It even can hover backwards up to 35 knots.

For touchdowns, the surface can have an uphill or downhill slant of as much as 15°, which is well above the limit of most helicopters. Single engine performance is relatively poor. Single engine hover is possible under only a very few conditions.

During conversion from hovering to conventional flight, there is a tendency to lose lift and sink, requiring the pilot to add power. But this is normal on VTOL aircraft. In summary, it can be said that through careful design of an extremely complex aircraft, flying the XV-15 and managing of all systems is straightforward.

With the nose up and full aft stick, level stalls in the clean configuration give a slight vibration at 110 knots. The aircraft will begin to sink, but there is no wing drop or other bad effects. Recovery is benign and quick. In helicopter autorotation mode, the best descent rate of 2,150 feet per minute is achieved at 75 knots. At 90 knots, the descent rate increases to 4,000 feet per minute. For final approaches, pilots quickly learned to use nacelle tilt angle instead of pitch inputs to control airspeed. It is different, but works very well.

The 1,400 pounds of fuel contained in the wings proved to allow for only about a 175 mile range. An auxiliary tank holding an additional 900 pounds eventually was added to the fuselage, which increased the range to about 325 miles.

In January 1981, hover tests were performed at Ames to evaluate downwash and noise. Tests showed that increased control activity was required as the aircraft enters ground effect. Downwash velocities were moderate at the sides and relatively high fore and aft. On a subsequent hover test at Bell when there was a light coating of snow on the ramp, it was noted that the downwash pushed the snow away from the aircraft, as expected. However, there was no "white-out," caused by snow being caught by the recirculating slipstream, as normally happens with a helicopter. Overall, it was determined that the aircraft's capabilities were not limited by gust sensitivity, aeroelastic stability, or downwash.

In March 1982, aircraft #1 made a demonstration tour of East Coast facilities, which included seven flight demonstrations at six different locations in eight days. One flight included a stop at the helipad at the Pentagon. While on the tour, the XV-15 flew 2,600 nautical miles and needed only routine daily preflight maintenance.

Following this East Coast tour, #1 was modified at Bell's Arlington facility to perform an electronics mission evaluation. Items added included an APR-39 radar warning system and chaff dispenser system. The aircraft departed for NAS China Lake in California in May, then on to Ft. Huachuca in Arizona in June, and finally on to San Diego for sea trials. Shipboard evaluations were performed aboard the amphibious assault ship USS Tripoli off the San Diego coast in July 1982. Fifty-four vertical landings and take-offs (of which five were STOL take-offs) were performed.

Other mission related evaluations included over-water rescue and simulated cargo lifting, which were demonstrated in May 1983, and simulated air-to-air refueling, which was performed in September 1984. By 1986, both aircraft had accumulated a total of 530 flight hours, made 1,500 transitions, and reached an altitude of 22,500 feet (while still maintaining an 800 foot per minute climb capability). In March 1990, #1 set numerous time to climb and sustained altitude records for this class of aircraft. These included a climb to 9,843 feet (3,000 meters) in 4.4 minutes and to 19,686 feet (6,000 meters) in 8.46 minutes, without even performing extensive climb tests to develop an optimal climb profile. It also sustained an altitude of 22,500 feet with a dummy payload of 2,200 pounds in addition to more than 1,000 pounds of test instrumentation.

As of June 1990, XV-15 #1 was based at Bell Helicopter's Flight Research Center in Arlington, TX, for continuing engineering development. XV-15 #2 was based at Ames for continuing tilt-rotor research. The two aircraft had accumulated 825 hours.

It is worth noting that most research aircraft were flown by only a small select batch of test pilots. Bell, however, felt that in order to insure the success of a production tilt-rotor aircraft some day in the future, a wide range of pilots should have the opportunity to fly the XV-15 and provide their inputs. Thus, by 1990 the XV-15 was flown by over 185 pilots with widely varying experience and capability levels, including several low-time private pilots. Numerous admirals, generals, and at least one U.S. senator and one service secretary flew as guest pilots. Each flight consisted of a brief demonstration of helicopter, conversion, and airplane modes by a Bell test pilot. The guest pilot then took over the controls. After a few minutes of familiarization, he was talked through

One of the XV-15s shown in a captive ground test stand in 1980. (Bell Photo)

An artist's concept of a production aircraft based on the XV-15 rescuing pilots while under fire. Note the machine guns on the under fuselage. (Bell Photo)

an airplane stall, single engine operation, and conversion/ reconversion at altitude. They then return to the airport for several take-offs and landings, usually converting to airplane mode and back to helicopter mode each trip around the pattern. Guest pilots rated the XV-15 as easy or easier than a helicopter to hover. Conversion was unanimously said to be straightforward, and with a low workload. Handling qualities in airplane mode were excellent. Most also noted the low interior noise and smooth ride.

FAA test pilots also flew the XV-15 in order to evaluate its potential for certification of a civil tilt rotor aircraft. While they saw no technical reasons for not being effective in the civil role, they determined that a review of Part 25, which sets standards for large transport aircraft, and Part 29, which sets standards for helicopters, would be needed in order to establish appropriate certification criteria.

XV-15 #1 remained in service at Bell's flight research center, where it was used as a concept demonstrator and marketing tool for the V-22 Osprey that was by then being developed. It was flown regularly until August 1992, when it was damaged beyond economical repair. A mechanical failure in the control system caused the aircraft to roll over while it was hovering. The crew was not injured, but the wing and one nacelle sustained extensive damage. At the time of the incident, #1 had flown nearly 841 hours. The forward fuselage was salvaged and put to use as a simulator to help develop Bell's upcoming civil tilt rotor aircraft.

XV-15 #2 remained at Ames through the 1980s. In 1986, it was fitted with composite rotor blades built by Boeing Helicopter. Sporadic testing was accomplished through 1991, when it was stopped due to a problem with the blade cuff that resulted in an emergency landing. While the blade cuff was being re-designed, NASA decided to put the airframe down for a major airframe inspection that would be due soon, anyway. Unfortunately, program funds again ran out before the inspection could be completed. #2 would remain partially dis-

assembled until mid 1994. It had accumulated just over 281 hours.

With Bell anxious to resume tilt rotor development, they established a Memorandum of Agreement with NASA and the Army in 1994 which transferred XV-15 #2 to Bell and allowed them to return it to service at no cost to the government. The disassembled aircraft was shipped to Arlington, Texas, and the refurbishment and inspection began in mid 1994. The original metal rotor blades were put back on, and the aircraft resumed flight testing in March 1995.

Much of Bell's recent research has focused on reducing noise in order to make civil tilt rotor more acceptable for operating in crowded urban areas. Tests were being conducted to determine the major sources of noise. (The familiar "wap-wap-wap" sound of a helicopter comes from the rotor blade passing through its own wake.) Bell is looking at combinations of approach profile, nacelle angle, and various rotor tip designs to minimize this noise.

In its current configuration, the XV-15 has a Rockwell-Collins glass cockpit that features a large, daylight readable liquid-crystal display that shows all flight information. It also displays flap and nacelle positions. The NASA white with blue paint scheme was replaced to enhance the marketing appeal for the civil tilt rotor development. As of the end of 1998, the remaining XV-15 had accumulated a total of 530 flight hours and remains in service at Bell's Arlington facility to continue developing and refining Tilt Rotor technologies.

BELL/BOEING V-22 OSPREY TILT-ENGINE VTOL TRANSPORT (U.S.A.)

System: Built-from-scratch prototype
Manufacturer: Bell and Boeing
Sponser: US Marine Corps
Operating Period: Early 1980s-to-present
Mission: To develop a twin-engine Tilt Engine transport for military applications.
The Story:
The Bell company brought a long history of vertical-lift Tilt-Engine system development to this program when it combined with Boeing in the early 1980s.

The company had demonstrated such a system with the XV-3, proving that it was possible to turn wingtip rotors from

An early concept drawing of the V-22. (Bell Drawing)

The V-22 in vertical ascention with engines in near-vertical position. (Boeing Photo)

a vertical position to horizontal, thus providing the thrust to lift the vehicle off the ground and then transition to horizontal flight. The XV-3 was followed by the XV-15, which would closely resemble the V-22, with the XV-15 initially serving as the test vehicle for the V-22.

During the early 1980s, the new Bell-Boeing competed for the lucrative Joint Services Advanced Vertical Lift Aircraft (JVX). Then, on April 26, 1983, the joint team was awarded the contract to begin the preliminary design stage.

Initially called the Bell/Boeing Vertol JVX, the plane would later be given the V-22 Osprey designation, and later, special mission variations carrying the designations of MV-22, HV-22, and CV-22.

The then-ambitious 1983 plan called for 1,086 of the tri-service VTOLs to be produced with a total procurement cost of $25 billion, with the funding to be shared by the three services. The Navy was to assume the bulk of the obligation with 50 percent contribution.

The initial requirement called for the craft to support all the services and be capable of carrying 24 troops. There were many other possible missions mentioned for the revolutionary new VTOL, such as electronic warfare, amphibious assault, and special operations.

Unlike a helicopter, the V-22 doesn't have to be disassembled and transported in a transport aircraft. With a single refueling it will be able to fly 2,100 miles.

Power for the Osprey, which is about half the size of a C-130, comes from a pair of General Electric T64-GE-717 turboshaft engines, each rated at 6,150 horsepower. The power

is definitely needed, since the V-22 weighs in at about 55,000 pounds maximum take-off weight in a STOL configuration, and about 44,000 pounds in a pure VTOL mode. Its empty weight is just over 33,000 pounds.

The rotor diameter is a large 38 feet, with an overall fuselage length of about 57 feet. With the rotors in the take-off position, the height of the vehicle is just over 20 feet. The width of the vehicle, with the rotors turning, is almost 85 feet.

Those rotors, by the way, are capable of rotating through 97 degrees and are constructed of graphite/glass fiber. In fact, 33 percent of the weight of the Osprey is fabricated of that high-tech material.

The rotors also possess separate cyclic control swashplates for sideways flight and fore-and-aft control dur-

V-22 carrying a camouflage paint scheme. (Boeing Photo)

A pair of V-22s transition to horizontal flight. (Boeing Photo)

ing hover. Lateral attitude is maintained by differential rotor thrust. Once the V-22 achieves horizontal flight, the engines have assumed a position parallel to the fuselage.

The V-22 can even fly backwards! This maneuver is accomplished by tilting the rotors back past the 90-degree vertical position. Such a capability might someday provide a useful combat tactic.

Initial performance requirements for the Osprey were a cruising speed of 300 miles per hour with a range with a full payload of over 900 miles. And even though the V-22 carries a cargo designation, it does carry a nose-mounted 12.7mm multi-barrel machine gun.

Since the V-22 will be operating in a combat environment at least a portion of the time, the plane's design has taken the situation into account. There are redundant systems, along with armor, in critical areas. The tough composite materials also provide protection. Also, one engine can power both rotors if required because of a cross-coupling capability.

The full-scale development program began in May 1986 with the order for six prototypes. Numbers 1, 3, and 6 were constructed by Bell at Arlington, Texas, and numbers 2, 4, and 5 by Boeing at Wilmington, Delaware. The first flights for the new vehicles occured between 1988 and 1991.

In 1995, the production of test vehicles resumed with numbers 7 through 10 for vibration, propulsion, USMC, and mating evaluations. Final assembly of the first production V-22 was completed in late 1998.

In 1998, Bell also began a two-year fatigue program to determine if the V-22 can accomplish a design life of 10,000 hours, or about the same as 20 years of operation. The rigorous program would subject the C-22 test vehicle to 18 million different aircraft loads.

The testing planned to simulate 7,000 hours in the normal aircraft role and 3,000 in the VTOL mode. No structural damage at 4Gs at over 300 knots is allowed. It is probably the most rigorous test program ever carried out on a new aircraft.

The complete array of engine positions for the V-22. (Boeing Photo)

The V-22 has been under development since the mid-1980s. (Boeing Photo)

In early 1998, the V-22 accomplished another milestone with its first night test flight. During the test, the V-22 flew with a neutral density filter which covered the cockpit displays. This allowed the pilots, which were using night vision goggles, to operate the displays in the day mode.

Also included in the flight test program was to investigate its capabilities for carry external loads. In one test, the number eight prototype carried a five-ton sling load through the transition phase with a top speed of 230 knots. The test pilot indicated that even though he could feel the load oscillating beneath him, it didn't affect the craft's performance.

The testing showed that certain loads caused dramatic shifts in the center of gravity of the V-22, which in turn required a change in the attitude of the aircraft. It was determined that that requirement could be accomplished by changing the angle of the engine nacelles.

Other tests planned at the time included a sling weight of 15,000 pounds and a Humvee jeep, hopefully faster than could be accomplished by the MH-53E Super Stallion helicopter.

In 1997, 14 years since the long-forgotten JVX contract had been awarded for design and development, low-rate initial production was undertaken for 23 V-22s (called MV-22s) for the USMC. This event occurred after the Pentagon's Defense Acquisition Board (DAB) had earlier given the go-ahead for production of the V-22. The first delivery was scheduled for 1999, with IOC occurring in the 2001 time period.

The projected cost of these first operational V-22s had a unit cost of $32.2 million each, a figure that was down considerably than the $41.8 million projected during the full-scale development program. The plane has attracted international interest, with both Israel and Australia both showing interest in the unique VTOL. Up to 423 V-22s are desired by the Marines, but there will probably be up to 50 built for the USAF for special operations activities.

Bell is building the wing section, empennage, and powerplant nacelles, while Boeing has the responsibility for

With its patriotic paint scheme, the test versions of the V-22 were unmistakable. (Boeing Photo)

The red smoke under the V-22 helped determine wind flow under the aircraft. (Boeing Photo)

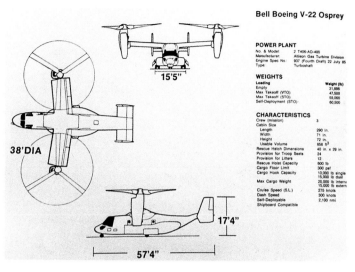

Bell Boeing V-22 Osprey

POWER PLANT
No. & Model: 2 T406-AD-400
Manufacturer: Allison Gas Turbine Division
Engine Spec No: 937 (Fourth Draft) 22 July 85
Type: Turboshaft

WEIGHTS

Loading	Weight (lb)
Empty	31,886
Max Takeoff (VTO)	47,500
Max Takeoff (STO)	55,000
Self-Deployment (STO)	60,500

CHARACTERISTICS

Crew (mission)	3
Cabin Size	
Length	290 in.
Width	71 in.
Height	72 in.
Usable Volume	858 in³
Rescue Hatch Dimensions	40 in. x 29 in.
Provision for Troop Seats	24
Provision for Litters	12
Rescue Hoist Capacity	600 lb
Cargo Floor Limit	300 psf
Cargo Hook Capacity	10,000 lb single
	15,000 lb dual
Max Cargo Weight	20,000 lb internal
	15,000 lb external
Cruise Speed (S.L.)	275 knots
Dash Speed	300 knots
Self-Deployable	2,100 nmi
Shipboard Compatible	

Data on the V-22. (Boeing Chart)

the fuselage and landing gear. Bell is also responsible for the final assembly and initial flight tests.

With cost always a huge consideration, serious consideration was given to faster Marine V-22 production for that very reason. It was assessed that cutting eight years off the production program, which was programmed to last through 2018, would save six billion dollars off the previously-estimated $42 billion.

Marine officials have indicated that they wanted to increase the production to 36 planes a year beyond 2004, up from the 30 now programmed. Whether that will occur remains to be seen in the late 1990s.

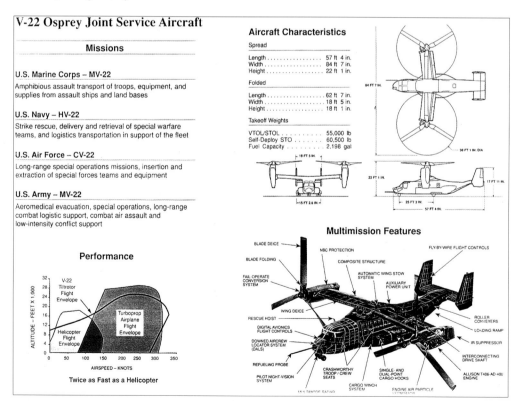

V-22 Osprey Joint Service Aircraft

Missions

U.S. Marine Corps – MV-22

Amphibious assault transport of troops, equipment, and supplies from assault ships and land bases

U.S. Navy – HV-22

Strike rescue, delivery and retrieval of special warfare teams, and logistics transportation in support of the fleet

U.S. Air Force – CV-22

Long-range special operations missions, insertion and extraction of special forces teams and equipment

U.S. Army – MV-22

Aeromedical evacuation, special operations, long-range combat logistic support, combat air assault and low-intensity conflict support

Performance

Twice as Fast as a Helicopter

Aircraft Characteristics

Spread

Length	57 ft 4 in.
Width	84 ft 7 in.
Height	22 ft 1 in.

Folded

Length	62 ft 7 in.
Width	18 ft 5 in.
Height	18 ft 1 in.

Takeoff Weights

VTOL/STOL	55,000 lb
Self-Deploy STO	60,500 lb
Fuel Capacity	2,198 gal

Multimission Features

All the aspects of the V-22 are addressed by this chart. Note the different names applied to the V-22 depending on the particular mission and using military service. (Boeing Chart)

The carrying capability of the V-22 was thoroughly tested. (Boeing Photo)

There is also a push for quicker acquisition of the V-22s for the U.S. Special Operations Command (AFSOC). The desire is to assume operational capability in 2004, with the first delivery in 2003. Even the Navy is looking at its own special version of the V-22 to bolster its capability to conduct covert operations.

The Osprey was also considered in the late 1990s for a search-and-rescue function. If approved, the Osprey could possibly replace the MH-60 helicopter, which now performs that function.

The projected capabilities of the V-22 over the helicopter point to its huge advantage. For example, the V-22 has a

The immense size of the V-22's propellers is vividly illustrated by this photo. (Boeing Photo)

Artist's concept of Bell Boeing 609 in Coast Guard livery. (Bell Photo)

24,000 foot altitude capability with a speed of about 270 knots. In comparison, the CH-46E helicopter shows a 9,000 foot altitude with a speed capability of about 125 knots.

It's for that performance advantage and its VTOL configuration that could allow the V-22 to replace a number of helicopter types in the future.

BELL BOEING 609 TILT-ROTOR VTOL(U.S.A.)
System: Built-from-scratch prototype
Manufacturer: Bell and Boeing
Sponser: Private company venture
Operating Period: Mid-1990s-to-present
Mission: To develop a commercial-application, nine-passenger, twin-engine, tilt-engine VTOL transport.
The Story:
Bell Boeing revealed in 1996 that studies had been in progress for some time for the development of a nine-passenger, Tilt-Engine VTOL passenger transport. The system was announced as being in the 14,000 pound weight class and carried a designation of D-600.

In November of that same year, the companies officially announced the joint venture to design, develop, and market the system, which would be called the Bell Boeing 609. The companies had earlier conducted extensive surveys in the civilian and commercial markets and concluded that a significant requirement existed for such a system. In fact, officials from both companies indicated that the need could possibly justify a thousand aircraft in the next two decades.

The companies feel that there are a number of possible missions, including emergency medical operations, law and drug enforcement, transport to and from off-shore oil rigs, search and rescue operations, and corporate executive transport.

A Bell Textron advertisement looking for engineers to work on the 609 project. (Bell Textron Advertisement)

It was also reported that the 609 could well be just the first of a whole family of such vehicles. In fact, there is a projected 620 version which could carry up to 20 passengers.

The seriousness of both Boeing and Bell for the program was accentuated by the fact that major subcontractors were selected during 1996. The prototype was also displayed publicly at the 1997 Paris Air Show, where it generated large interest. Following that show, the 609 was then displayed for a number of potential European customers.

The 609 will undoubtedly fill many commercial needs. (Bell Photo)

With its somewhat-normal appearance, when the engines are in the non-tilted position, the craft doesn't have the appearance of a Tilt-Engine vehicle. As such, it is perceived more as a standard twin-engine corporate jet instead of a revolutionary new aircraft development.

The 609 has also been proposed as an addition to the helicopter and fixed wing fleet of the U.S. Coast Guard. In fact, the 609 was evaluated in 1998 by the Coast Guard as a part of its "Deepwater" initiative, which would involve the upgrading of Coast Guard aircraft for the 21st century. An initial assessment of 60-90 609s might be required for this mission if it were to be selected.

Planning for the actual production of the 609 was well underway by 1997. Officials explained that the fuselage would be produced at Boeing's Vertol facility near Philadelphia, while the wing, as well as final assembly, would be accomplished at Bell's Fort Worth operation.

Plans during the late 1998 time period called for certification in both the United States and Europe during the first quarter of 2001, with deliveries planned to commence shortly thereafter. As of late 1998, there had been orders for some 60 planes. At that time, about one-third of the planes would support natural resource companies, another third to air-taxi companies, while the final third would go to corporate customers. This wide range of uses points to the flexibility of the Tilt-Wing model.

The 609 incorporates a unique advantage in that it will be able to land on heliports atop large buildings in metropolitan areas, a la a helicopter. But once in flight, and following the transition from vertical to horizontal flight, the 609 can assure the high-speed characteristics of a corporate jet, along with the sleek looks of that type of machine.

The model is projected to be in the $8-to-10 million dollar unit cost range. It should be noted that figure is considerably higher than a conventional helicopter, so the speed and payload advantages of this system must be evaluated by the prospective customer.

An extensive flight test program was planned with the construction of four test planes. The fourth vehicle would be used by the contractors as a demonstration vehicle. The goal of the flight test program was to firmly prove every aspect of the concept, looking at every part of the vehicle from a performance and cost point-of-view.

The projected capabilities of the 609 are impressive, with a cruise speed of 275 knots, a service ceiling of 25,000 feet, and a maximum range of 750 miles. All of those capabilities greatly exceed those of existing helicopters, making the 609 a very attractive alternative.

It has been noted that the 609 could cover about five times the area covered by a helicopter in a search and rescue operation, along with transit times to and from a destination being reduced as great as fifty percent as compared to pure rotary-wing aircraft.

Compared to the V-22 military system, the 609 is a very small system, but the concept is basically the same with the same Tilt-Rotor mechanism.

The 609 design incorporates a T tail, which raises the horizontal member above the rotor wake and reduces their pitching moments during the transition portion of the flight. The relatively small-span wing size was determined by the need for it to hold all the fuel for center of gravity and safety considerations. The engines are cross-shafted, using composite materials, which will enable the craft to stay aloft in case of an engine failure. Also, in the case of cross-shaft failure, there is a screwjack arrangement which allows the rotors to be tilted into a vertical helicopter-style mode.

The flying controls are high-tech, with a Lear Astronics triplex digital fly-by-wire system. The T tail has no rudder, but is equipped with conventional elevators and two-segment trailing edge flaperons.

Looking at the structure, composites again play heavily in the skinning of the wing and fuselage. The fuselage is constructed in three main sections, the nose, center section, and tail. The upper and lower wing surfaces were each produced as a single piece. In all, the 609 weighs in at a minimal 14,000 pounds.

Due to its large use of composites, the production procedure for the 609 will be modified. Spools of carbon fiber material will be fed to computer-controlled machines. There, the material is coated with epoxy and then is laid on a mold in long crisscrossing bands.

The procedure came about from the V-22 experience when Boeing found out that it could use such skin panels to replace numerous smaller panels, thus eliminating the need for hundreds of fasteners.

Landing the 609 is accomplished on a retractable tricycle landing gear, with twin nose wheels and a single wheel on each rear fuselage unit.

Artist's concept of Bell 609 in forward flight. (Bell Photo)

Power is provided by a pair of 1,850 horsepower Pratt & Whitney Canada PT6C turboshaft engines, each of which is installed in the wingtip tilting nacelle. Each engine drives a large three bladed propeller. The unique use of the engines required some significant modifications, including a modified oiling system with dual pumps in order to provide sufficient oil pressure when operating in the straight-up mode.

The 609 carries a crew of two, located side-by-side on the flight deck. The cockpit features three large electronic displays, built by Rockwell Collins. The standard version is also capable of carrying a passenger load in air conditioned and pressurized comfort.

CURTISS-WRIGHT X-100 TILT PROP VTOL (U.S.A.)
System: Built-from-scratch research aircraft
Manufacturer: Curtiss-Wright
Sponsor: Curtiss-Wright
Time Period: 1958 - 1961
Mission: Technology demonstrator to verify utility of radial lift concept
The Story:
The Curtiss-Wright X-100 was, by intent, a short-lived research program used to demonstrate the radial lift concept being developed for the Curtiss-Wright M-200 aircraft, which eventually became the X-19. As such, the X-100 had only two purposes. One was to test the radial lift force concept and the gimbaled nacelles needed for a tilt prop configuration. The other was to test the glass fiber propellers needed to enable the use of radial lift. The X-100 was successful in that it verified the characteristics of the props, and produced data on noise, vibration, downwash, ground effect, stability, and control, and the piloting techniques needed to hover and transition a tilt prop VTOL.

The airframe itself was simple and straightforward. It was all aluminum construction except for the aft half of the fuselage starting just aft of the cockpit, which was fabric covered. Design and fabrication was completed in just over a year, using what is now called concurrent design and construction. Basically, this means the design, fabrication, testing, and documentation of components occurred in a continuous and ongoing process.

The X-100's basic configuration was a high wing with T-tail and conventional landing gear. The fuselage was 28.3 feet long and consisted of welded tube construction. The small, shoulder mounted wing, with a span of only 16 feet and an area of 22.5 square feet, had a very heavy wing loading of 170 pounds per square foot. This was intentional so that the radial force produced by props could be demonstrated beyond any doubt. A three bladed, 10 foot diameter propeller was mounted on a gimbaled nacelle at each wing tip, and the two props rotated in opposite directions to balance yawing moments. Gross weight was 3,500 pounds. The two-place cockpit had side by side seating. A rectangular engine intake was just behind the cockpit on the top of the fuselage. The T-tail, with an area of 23 square feet, was chosen to keep the

stabilizer out of the props' slipstream, giving the X-100 an overall height of 10.75 feet.

To control pitch and yaw during hover, the engine exhaust was ducted out the rear of the fuselage and could be vectored up or down, and left or right by a device called a jetivator. This device was chosen for simplicity and to reduce weight. As a control effector, the jetivator could produce 140 pounds of pitch force and 40 pounds of yaw force. Roll control during hover was by differential control of the prop pitch. Altitude control was by means of the throttle.

The X-100 originally had a very rigid conventional landing gear. It eventually was replaced by tricycle gear to improve landing characteristics.

The props were made of foamed plastic molded onto a metal shank and covered with a layer of urethane elastomer. Blades were wide and had a considerable amount of twist to maximize the amount of radial lift. The relatively low tip speed of 650 feet per second in hover minimized the amount of noise.

The nacelles could rotate from pointing vertical to 12 degrees above horizontal. They rotated forward slowly at 2 degrees per second to allow the pilot to adjust power and attitude in order to prevent settling during the transition to cruise flight. They rotated back to vertical at 5 degrees per second.

Keeping with Curtiss-Wright's philosophy of building all major systems, management wanted to use a Curtiss-Wright developed or license-built engine. However, program managers hated to use internal funds to design a new engine. Originally, they considered a French Turbomecca Artouse, producing about 400 horsepower. It was abandoned during the design phase in favor of a Lycoming YT-53-L-1 turboshaft engine of either 650 or 825 horsepower (references varied). The U.S. Army, who had numerous YT-53-L-1 engines on hand, loaned two of them for the X-100 project. In exchange, Curtiss-Wright agreed to write a report on the engines' performance. The fuselage had to be enlarged to accommodate the Lycoming engine, but this was not a significant expense considering the money saved with the engines loaned from the Army.

The design process started in February 1958. The first prop assembly was fabricated soon after and tested in NASA's

One of the first Tilt-Prop designs was the Curtiss-Wright X-100. (Curtiss-Wright Photo)

40x80 foot wind tunnel starting in October 1958. It was run through the expected power operating range and at shaft angles from 0 to 90 degrees. These tests demonstrated that thrust in hover would be 10 percent below prediction (which was later confirmed in flight test), but there was enough excess thrust that the 10 percent loss was acceptable. Preliminary wind tunnel tests using models were performed at the Massachusetts Institute of Technology, then a powered model was run in the wind tunnel to determine stability and control characteristics. Tests using the actual aircraft eventually were run in NASA's 40x80 foot wind tunnel. Reasonable correlation was achieved between wind tunnel tests and flight tests.

Roll out of the completed X-100 occurred on December 22, 1958, at Curtiss-Wright's Caldwell, NJ, facility, and the first engine run was performed on January 14, 1959. The first hover was made in a tether rig on April 20, 1959, to verify control power and feel. The first free hover test was performed on September 12, 1959, but extra weight was added to prevent the X-100 from getting out of ground effect. After three days of testing, hover flights of up to 20 minutes duration were made. The outflow velocities of air along the ground at various distances from the hovering X-100 proved to be similar to those of a helicopter of similar weight.

Hover proved to be difficult for several reasons. The pitch and yaw thrust provided by the jetivator was insufficient. Roll and yaw motions tended to couple together. The throttle was used to maintain height during hover, and lags in the engine response further compounded the problem. The X-100 also tended to weathervane into the wind during hover due to yaw moments produced by the wind hitting the props at an angle, which the jetivator's yaw control could not counter adequately.

Steady hovering was possible only up to about 13 feet, while the X-100 was in ground effect. If maximum power was applied on the ground, the X-100 would rise to about 25 feet, settle to within a few feet of the ground, then rise again and repeat several cycles of a slowly damped oscillation in height.

The first transition from hover to forward flight was made on April 13, 1960. The transition itself was fairly simple. The pilot pressed a switch on top of the control stick to rotate the nacelles forward, then added power to prevent sinking. As the nacelles rotated, the X-100 picked up 20 miles per hour of airspeed for each 10 degree of tilt. The conversion was done at 3 to 5 feet of altitude and very slowly to prevent settling to the ground. The nacelles were rotated down in 5 degree increments until a speed of 180 miles per hour was achieved with the nacelle at a 15 degrees from horizontal. In general, low speed flight characteristics were unsatisfactory, especially in gusty air. The X-100 demonstrated a high pitch-up moment as forward speed increased, but directional stability, roll control, and longitudinal maneuvering were good in flight at 70 miles per hour. At low altitude and up to 70 miles per hour, the X-100 was notably free of mechanical vibration. Testing at altitude showed the aircraft to be buffet and stall free over a large range of angles of attack.

STOL take-offs and landings were made with nacelle angle at 20 degrees from the vertical. Directional control was poor and was a problem during ground run. Smooth touch-downs also were difficult, much of the problem being caused by the landing gear being too stiff. There was considerable bouncing even at low touchdown speeds.

The basic test program was completed on July 21, 1960. Test pilots from NASA visited Curtiss-Wright and flew the X-100 on August 12 and 13. This concluded test operations for the X-100 at the Caldwell, NJ, facility.

With Curtiss-Wright satisfied that they had proven the radial lift concept, they turned the X-100 over to NASA. It was shipped to the Langley Research Center in October of 1960. However, it was not really used to test flight characteristics, but to study the effects of downwash on several types of ground surfaces, such as snow, grass, pavement, and packed dirt. NASA also studied problems associated with visibility in snow during hover and slow forward transition. This role, too, was finished by October 1961.

Following the completion of tests at Langley, the X-100 was donated to the Smithsonian Institution. During its short career, the X-100 completed a total of 14 hours flying time and 220 hours of ground engine run time. It successfully proved the radial force concept, the feasibility of tilting nacelles, and low noise levels. It successfully showed the value of using a demonstrator aircraft to highlight technical concerns quickly and to reduce risks for higher cost development programs.

CURTISS-WRIGHT X-19 TILT PROP VTOL (U.S.A.)

System: Built-from-scratch VTOL research aircraft
Manufacturer: Curtiss-Wright
Sponsor: U.S. Air Force
Time Period: 1958 - 1965
Mission: Tilt Prop technology demonstration and development of radial lift concept
The Story:
The Curtiss-Wright X-19 started life as a commercial venture to develop a small, 4-passenger, executive VTOL aircraft that would have good high speed performance. Funded initially by Curtiss-Wright, it was designated the M-200 (references

Tilt-Prop X-19 in lift-off configuration. (Air Force Museum Photo)

vary, some calling it the X-200). The M-200 was a very innovative aircraft with which Curtiss-Wright would attempt to reenter the aircraft business. Unfortunately, it proved to be their last try.

The Curtiss-Wright Company was formed from the merger of the companies started by aviation pioneers Glenn Curtiss and the Wright Brothers. Curtiss-Wright was a major aircraft manufacturer through World War II, but after the war they never won another production contract and gradually entered other markets. By 1952, they even disbanded their aircraft design and manufacturing division.

However, Curtiss-Wright remained a giant in the propeller business. In 1958, Curtiss-Wright engineers envisioned the idea of using a propeller's radial lift to generate the lift required to fly at low forward speeds and up through transition. The prop not only would propel the aircraft forward, but also would provide lift during forward flight. This would free the aircraft from having a wing sized for low speed flight. With a practical application of this concept, Curtiss-Wright hoped to reenter the aircraft market.

To explain radial lift simply, think of a propeller turning on a shaft pointed straight into the air flow. In this case, each blade of the propeller strikes the air flow in exactly the same way. However, when the propeller shaft is at some angle relative to the airflow, the blades are not loaded evenly, as it would be if moving straight into the wind. This is because the blade moving down is at a higher angle of attack than the rest of the blades. The net result is a force, called the radial lift, which is perpendicular to the shaft and pointing upward. Obviously, a propeller produces a large amount of thrust in the direction parallel to its rotating shaft. Radial lift is in addition to this thrust force. All propellers produce radial lift that in general is thought of adversely. But, in theory, if the propeller is designed properly, radial lift could provide enough lift to completely support an aircraft. The amount of radial lift that can be produced is a function mostly of the blade width and twist, accounting for the wide, highly twisted shape of the X-19's props.

To confirm the radial lift principal, Curtiss-Wright built and flight tested an experimental VTOL aircraft called the X-100. With solid test results confirming their theories regarding the potential for a practical application of radial lift, Curtiss-Wright pressed on with the design of the M-200. In theory, the wing could now be sized for cruise flight, since the radial lift produced by the props would provide a significant portion of the total lift needed at low speed. Numerous design trade-off studies were performed which ultimately led to the X-19's tandem wing configuration with a tilting propeller at each wing tip.

It is interesting to note that the X-19 was built from the propeller designers' point of view. As a propeller manufacturer, Curtiss-Wright long had observed that in most cases the aircraft designers did not consider propeller requirements until long after an aircraft's configuration was frozen. The consequence was that the prop had to fit into other aspects of the aircraft design that already were fixed, with the resulting propeller design often being far from optimum. Thus, the X-19 started with what Curtiss-Wright engineers felt was the optimal design for a propeller for a VTOL aircraft and then built the aircraft around it. The propellers had a somewhat awkward appearance, being very wide with a large amount of twist. This allowed them to maximize the amount of radial lift.

At the start of the program, Curtiss-Wright funded the entire effort, with no interest in any government support. Two prototypes were being built when new management at Curtiss-Wright decided they no longer wanted to invest company research funds. They decided to offer the two aircraft to the Tri-Service VTOL Program, a joint Air Force, Army, and Navy program office tasked with developing VTOL technologies for military needs. The Tri-Service Program also was developing the XC-142 and the X-22. It took eighteen months of trying until the Air Force agreed to buy the aircraft. Curtiss-Wright's interest in the effort was fading rapidly, and the Air Force's decision to buy the M-200s came on the morning that the board of directors was meeting to make a decision on con-

Four props in the vertical position provided for a smooth VTOL lift-off for the X-19. (Curtiss-Wright Photo)

tinuing the program. When the Tri-Service Program bought the X-19s, the two prototypes already were 55 percent and 35 percent complete, with a substantial investment already having been made by Curtiss-Wright.

The original X-19 design, as the M-200, was not built to any specific mission requirement. The goal was to fly as fast and far as possible and to be economically competitive with conventional executive transports that were becoming very popular in the 1960s. Analysis of the design showed that the M-200 could achieve a range of 900 to 1,150 miles with a maximum level speed of 400 knots at 16,000 feet. It was to conform to FAA regulations, have all weather flying capability, low noise level, and be free of vibration. Conventional aircraft construction techniques and materials were used.

The X-19's basic configuration was an all-metal, monocoque fuselage, with two shoulder-mounted tandem wings. A nacelle at each wing tip could rotate from pointing vertically for take off and landing to pointing horizontally for cruise. The wide, specially-designed propeller was mounted in front of each nacelle. Two turboshaft engines housed in the rear fuselage powered the four props. The fully hydraulic tricycle landing gear retracted completely into the fuselage. A large vertical tail was required because the props were located relatively close to the fuselage. Total height was just over 17 feet.

The original fuselage was 41 feet long, but the cabin area was small…only 4 feet high, 4.5 feet wide, and 8 feet long. The passenger compartment was intended for four passengers, or 1,000 pounds of cargo. The cabin was pressurized to 16,000 feet. Maintaining pressure to a greater altitude would have added too much weight. All fuel was stored in the fuselage, aft of the passenger/cargo area. There were two 229 gallon tanks and one 261 gallon tank.

The front wing had a 20 foot span with a narrow chord, while the rear wing had a 21 foot span with a greater chord. The rear wing had almost twice the area of the front wing. The wings had no incidence, dihedral, or sweepback. The front wing incorporated full span flaps, while the rear wing had inboard ailerons and outboard elevators, the ailerons being slightly larger than the elevators. The flaps on the front wing were directly coupled to the nacelle tilt angle, and the pilot could not control them independently. At hover, the flaps and elevators drooped to their full extension of 60 degrees to decrease the amount of wing area that was in the prop downwash. The location of the props on wing tips, however, resulted in the loss of 7 to 9 percent of lift due to wing interference.

Curtiss-Wright intended originally to use four Wankel rotary engines rated at 580 horsepower each, with the capability to still operate vertically when one failed. The Wankel engine was attractive because it could hold sea level power up to 20,000 feet. Curtiss-Wright wanted to control the manufacturing of all major components that would go into the aircraft and purchased rights to the Wankel engine for North America, as well as aviation rights worldwide. However, they could not get a favorable licensing agreement from the manufacturer. Curtiss-Wright eventually abandoned the Wankel in favor of two Lycoming T-55-5 turboshaft engines of 2,200 horsepower each (although some sources stated T-55-L-7 engines of 2,650 horsepower each). This more than doubled the total power, but retained the ability to operate with one engine failed, including performing vertical take-offs and landings. Switching from four engines to two also simplified the design by decreasing the engine interconnects, and changing to turboshaft engines eliminated the need for engine cooling. Power from the engines was distributed to the props by means of three-inch diameter drive shafts and seven gear boxes. The gear boxes consisted of one engine coupling box that coupled the two engines so that either could power the entire system, two T-boxes to transfer power from the fuselage shafts into the wing shafts, and four nacelle tilt gear boxes. The exhaust pipe from each engine joined in the fuselage so that only one pipe exited the rear of the aircraft.

The 13 foot diameter props had a unique appearance, as previously discussed, being optimized for the production of radial lift. Their construction consisted of a steel shank, foam core, and fiberglass shell. This innovative, light-weight construction was needed because conventional metal props with the required design would have been too heavy. The paddle-wheel shape allowed them to produce about 6 pounds of thrust per horsepower, a relatively large amount. Low prop noise was obtained because the maximum tip speed of 820 feet per second was well below sonic speed.

The cockpit accommodated two pilots seated side by side. The aircraft could be flown from either seat, but the pilot-in-command seat was on the right, as in a helicopter. The cockpit resembled that of a conventional aircraft, with the cockpit visibility being very limited for a VTOL aircraft. The high instrument panel also restricted vision over the nose. Two one-foot square windows were located near the pilots' feet, but were too small to be effective and later were modified into ram air inlets to provide additional cooling. Each pilot had a conventional stick, rudder pedal, and two throttles. Nose wheel steering was by means of a hand tiller controlled by the copilot.

To minimize the clutter on the instrument panel, there was a single oil temperature gauge and single oil pressure gauge for all nine gear boxes. The pilot could select which component to monitor by rotating a 9-way selector switch to the desired component. To signal a problem, each gear box had a single warning light that indicated either high or low oil pressure, or the presence of metallic chips in the oil. If the light illuminated, the pilot had to rotate the 9-way switch to see the specific problem.

When rotated to the vertical positions, the front nacelles rotated past vertical, to 97 degrees, meaning the thrust actually pointed slightly forward. The rear nacelles rotated only to 82 degrees with their thrust pointing slightly rearward. A tilt button on the control stick caused all nacelles to tilt together. Initially, the nacelle tilt rate was mechanized at 5 degrees per second. It was soon realized that this would have required a deceleration from 50 knots to hover in less than 5 seconds, resulting in .5g of longitudinal acceleration. This was too much, so the rate was reduced to 1 degree per second, resulting in a more reasonable .2g. Two independent hydraulic systems controlled the nacelle rotation and prop blade angles. If the automatic system failed, the pilot could rotate them manually using a hand crank. The crank required 570 turns to move the nacelles from end to end, and eventually was replaced with a motor.

The Tri-Service Program required a few changes to convert the M-200 into the X-19. North American Aviation LW-2B ejection seats, which used a rocket catapult mechanism and ballistically deployed parachutes, were added. These seats could operate from zero altitude and zero airspeed up to 50,000 feet and deploy fully within .5 second from seat firing. The seats ejected through the canopy. The fuselage was lengthened by three feet so that two more passenger seats could be added, bringing the total personnel load up to two pilots and six passengers. This increased the fuselage length to 44.4 feet. The cabin door was enlarged to 42 inches high and 46 inches long, and a rescue hoist was added. The resulting aircraft was a bit long for a six passenger aircraft, but this was attributed mostly to the engines and fuel tanks being located in the fuselage. The only difference between the two prototypes was that the second had better instrumentation and data recording, and also had a dummy refueling probe installed on the nose to evaluate probe and drogue in-flight refueling.

Weight growth after adding the ejection seats and stretching the fuselage resulted in the useful load being reduced to 410 pounds. This allowed for only one pilot and fuel for 10 minutes of hovering. With two pilots, the hovering time was cut to 1 minute. However, there was still 600 pounds of unusable fuel because of the location of the fuel pumps inside the tanks. The pump position was changed to improve the amount of usable fuel.

The X-19 was statically unstable in hover and in pitch and roll at low speed, mandating the addition of stability augmentation in both of these axes. Control by the augmentation system was limited to 30 percent of the pilots' control authority, so the pilot could override it if necessary. No stability augmentation was needed for yaw. The system initially mechanized rate feedback, but initial testing showed this to be of little use to pilot. The system was changed to a rate plus integral of rate, which improved control for hover and low speed flight. The X-19 was test flown with the stability augmentation turned off. Although it was controllable by the pilot, the workload was unacceptable.

The X-19's "Achilles' Heel" proved to be its gear boxes. All gear boxes were designed to absorb the power required to lift the aircraft at a maximum weight of 12,300 pounds. Great difficulty was encountered in qualifying them, and very low life limits were established in order to conduct flight tests.

The engine coupling gear box, which combined the power output of the two engines, was capable of absorbing 2,900 horsepower with both engines running, or 2,500 horsepower from only one engine. With each engine producing 2,200 horsepower (or 2,650 horsepower, depending on the reference), the gear box obviously could not absorb full power from both engines. In addition, it was limited to 50 hours of operation. Likewise, the nacelle gear boxes had estimated lives of 13-14 hours and were thus limited to 5 hours of operation. The problem was felt to be improper heat treating of the gears.

Wind tunnel tests of the initial design indicated an unacceptably high amount of interference drag at the junctions of the wing/fuselage and wing/nacelle. Drag clean up measures were taken, and more wind tunnel tests were run to confirm the improvement. The rear wing had significant lift loss due to downwash from the front wing, but knowing this, it was compensated for in the basic design. Airframe structural tests revealed no major problems. Full scale static propeller tests were performed to evaluate and reduce download losses on the wing at hover.

The intended design empty weight was 8,000 pounds with a gross take-off weight of 12,300 pounds. However, the empty weight grew to 10,675 pounds and the gross take-off weight to 13,660 pounds by the time the aircraft was built, thus reducing the useful load and placing greater stress on the gear boxes. Most of the weight growth came from the fuselage, wings, and power transmission system. It is safe to assume that Curtiss-Wright engineers started their fuselage and wing designs using standard practices for a conventionally configured aircraft, but stress analyses indicated weaknesses peculiar to the tandem design and prop locations. The fuselage had to be heftier than originally expected to handle the loads imposed by the landing gear location and the tandem wing configuration. The wings also had to be heftier than normal because during hover, all lift came from the props at the wing tips, rather than the lift being distributed along the wing as on a conventional aircraft. Also, the wings had to be very stiff so as not to transmit any propeller vibration. The shafting and gear boxes also proved to be heavier than predicted.

Altitude was controlled during hover by the throttle, and precise control was difficult because of a lag of nearly one second in the engine response time. Varying the pitches of the four props controlled pitch and roll. A unique prop rotation scheme was used to maximize propeller torque for yaw control…the props on opposite corners turned in the same direction. A yawing moment resulted from the blade angle being increased on one corner and decreased on the opposite corner. The pilot didn't control the blade angles directly, but used the rudder pedals for yaw and the stick for pitch and roll inputs. A mixer in the flight control system automatically controlled the actual commands to the individual props.

To initiate a transition from hover to horizontal flight, the pilot pressed the tilt button on the stick to start the nacelles rotating, then added throttle to increase lift, accelerate, and maintain altitude or climb. As altitude and speed increased, the pilot continued to lower the nacelles until reaching 160 knots, at which time the transition was complete and the X-19 would fly like a conventional aircraft. Throughout the transition, the mixer continuously faded out prop control and faded in conventional control surface control as the airspeed increased. To reverse the transition, the process was reversed.

Roll out of the first prototype, tail number 62-12197, occurred on July 23, 1963. The flight test approach was to demonstrate hover, transition, and finally forward flight. Curtiss-Wright pilots would fly the initial flights, demonstrating hover and transition. After that, a Tri-Service test team would take over.

The first flight was performed on November 20, 1963, at Curtiss-Wright's facility at Caldwell, NJ. The X-19 lifted off for only a few seconds in hover before settling and collapsing a main gear because of side loads. Although the damage was minor, it was seven months until the X-19 flew again.

After this first accident, the Air Force and Curtiss-Wright debated over where flights should be performed. The Air Force wanted to send the X-19 to Edwards AFB. Curtiss-Wright argued that supporting the test program at Edwards would be difficult, and that in an emergency, the X-19 would land vertically, making the long runway and large dry lake bed unnecessary and a waste of taxpayers' money. The Air Force finally agreed that flights up through transition would be done at Caldwell, NJ, but the remainder of test flights would be performed at Edwards.

Flight testing resumed on June 26, 1964. The X-19 made numerous hovers, most only a few seconds in duration. By August 7, it had flown on twelve different days and accumulated one hour and 37 minutes, accomplishing most of the objectives planned for its first eight hours of hover testing. The X-19 demonstrated spot turns, lateral translations at

speeds up to 15 knots, rearward flight at 10 knots, forward flight at 20 knots, and 50 take-offs and landings. This may sound meager compared to testing fixed wing aircraft, but was consistent with the way helicopters were tested.

During these tests, the X-19 proved difficult to control, requiring excessive pilot workload. Various combinations of stick breakout and gradient forces were tried, but produced no improvement. However, the pilots were improving their skills at such a fast rate as they gained experience that it was difficult to determine if improved performance was from increased experience or the control system improvements.

On the 21st flight, the stability augmentation system was turned on, which made hovering much easier. Flight speeds up to 100 miles per hour were obtained with good flight characteristics up to this speed.

On November 12, the X-19 experienced a full pitch hardover caused by the stability augmentation system. The pilot retained control because of the limited authority of the stability augmentation system and deactivated it. On the next day, they experienced a roll hardover. The cause of these incidents never was determined conclusively.

On December 4, 1964, the X-19 suffered damage from loose cinders on the runway, which had just been resurfaced. Both engines and the leading edges on all four props were damaged. On January 31, 1965, one prop failed, which forced suspension of further testing for six months.

While repairs were being made, it was decided that testing should be moved to the Federal Aviation Administration's National Aviation Facilities Experimental Center (NAFEC) near Atlantic City, NJ. On July 31, 1965, the X-19 hovered for an air worthiness test after six months of down time, following which the props were removed and the aircraft was shipped to NAFEC.

The X-19 was reassembled quickly and soon was flying again at NAFEC. Throughout August 1965, it made numerous high speed hovers going more and more into the transition. At this point, the Air Force test team joined the program. The Air Force test pilot quickly determined that the remaining control problems during hover were caused by excessive and uneven hysteresis in the control stick. Hysteresis is the tendency of the stick and control surfaces not to return to their original position when the pilot moves the stick then lets go (it is caused by friction between all the moving parts and stretching of control cables). Pitch hysteresis was found to be 4 to 8 percent of total displacement, and lateral hysteresis was 12 to 25 percent. Pilots had to make much bigger roll inputs than pitch inputs, resulting in poor control harmony and the pilot getting out of phase with the aircraft's motions. This problem was corrected, and hovering became much easier.

The first full transition was planned for August 25, 1965, on flight number 50. As the X-19 climbed and accelerated, the nacelles tilted as far as 65 degrees from vertical as the aircraft reached 80 knots at 1,300 feet. At this point, the temperature warning lights for left rear nacelle gear box and aft T-box illuminated. The pilots terminated the test and began a

X-19 under construction at the Curtiss-Wright produciton facility. (Curtiss-Wright Photo)

return to the airport. An immediate, emergency landing did not appear warranted, and the pilots planned a normal approach based on the recommendation from the ground support team. As they circled to line up with the preferred runway, a very high frequency vibration and a low frequency random shake began. Lateral control began to deteriorate. With the X-19 pointed toward a wooded area at low altitude and fearing that they would not clear the trees, the copilot jammed the throttles forward. As they climbed and accelerated, some control was regained and the vibrations smoothed out. At 400 feet, the left rear prop snapped off, and the X-19 rolled to the left and pitched up. This was followed promptly by the separation of the left front prop, then the two right props. The pilots bailed out inverted, their parachutes deploying fully within 2 seconds at an altitude of 230 feet. The aircraft crashed into a nearby swampy area and was totally destroyed. The pilots suffered only minor injuries, mostly cuts from ejecting through the canopy. The time from loss of props to ejection was only 2.5 seconds. The cause of the crash was attributed to the copilot's applying full power. The drive system could not absorb the 4,400 horsepower being generated, resulting in the failure of the prop gear box. Total flight time for the flight was only seven minutes.

The crash itself did not end the X-19 program. The Air Force wanted to continue with testing the second prototype, which was nearing completion, but wanted to switch to a fixed price contract. Curtiss-Wright did not like that, because they would now take all financial risk. Their management also saw no future business in a commercial VTOL transport, even if the X-19 ultimately proved successful. They refused the Air Force's offer. Unable to come to any agreement, the program ended in December 1965.

The X-19 completed 129.4 hours ground running time and flew a total of 3.85 hours. The second X-19 never flew and eventually was cut up for scrap.

An interesting side note can be made regarding the data that was available on the X-19. Typically, during the life of a program, numerous technical papers and journal articles are published. But, when a program terminates and the people scatter to other projects, press releases stop and little effort is made to document and preserve final results and lessons learned. If it exists, it usually is stashed on peoples' shelves, eventually shoved in boxes relegated to an attic or basement, and seldom entered into the technical library system. This hinders future researchers who try to learn from past successes and failures. The X-19 was a refreshing departure from this trend. Numerous papers were written and preserved, documenting both the accomplishments and lessons learned. Indeed, the length of this chapter and even the few conflicts in information is testimony to the vast amount of information that was preserved and found readily. It is especially noteworthy that some of these documents were written nearly twenty years after the program was terminated, indicating that the Curtiss-Wright engineers who worked on the X-19 program believed in the technology they developed and wanted their work to be remembered.

5
Vectored Thrust VTOL Systems

The Vectored Thrust VTOL principal of redirecting the direction of engine thrust could be considered one of the most-basic VTOL techniques of the many that have been considered and tested.

Initially, the power requirements are considerably less since the entire powerplant does not have to be moved, only the whole, or a part of, the exhaust nozzle. A number of variations of the basic theme have been examined, and a number of them have been successful, resulting in their use in a number of operational aircraft.

Such devices as louvered nozzles, vectored nozzles, an array of so-called diverter vanes, thrust valves, and others have been tested through the years. It should be noted in all these concepts, a power source to move either the nozzles themselves, or the device inserted into the thrust stream is required.

The normal transition technique of this VTOL type is to accomplish the take-off with the thrust stream pointed close to straight down. Then, once the desired altitude is acquired, the thrust is slowly diverted to a horizontal direction along the centerline of the fuselage.

As a general rule, Vectored Thrust VTOL aircraft usually have a smaller wing area than conventional aircraft, since the propulsion system may provide lift during horizontal flight. At high speeds at low altitudes, the fuselage and tail surfaces might be able to provide enough lift to maintain a constant altitude. Since sizable-area wings are not needed for the high speed take-offs and landings of conventional aircraft, the smaller wing concept with this technique appears very advantageous.

Inherent with this system is the fact that the direction of thrust is normally applied in the optimum direction.

It should also be noted that in the Vectored Thrust concept, it was sometimes necessary to augment the hovering mode with a control function. This was sometimes accomplished with small jet pulse engines, usually located on the wingtips, which helped maintain the wing in a level attitude.

Over the years, the concept has been looked at seriously by the United States, United Kingdom, and the Soviet Union with a number of systems evolving.

Possibly the first Vectored Thrust concept was the so-called Dornier Aerodyne E.1. The strange-looking craft had a shrouded propeller, and the slipstream was deflected down-ward by cascade-like vanes for VTOL take-off and landing. The E.1 made its first flight in 1972 with a number of successful unmanned flights during the 1970s.

Another interesting Vectored Thrust project was the VZ-9V Avrocar, which would be canceled before any production was undertaken.

The VZ-9V, built by the British Avro Aircraft company, was an attempt to take advantage of the saucer shape. The craft was circular in shape, using a central fan which was powered by a trio of turbojet engines. The thrust was directly down during the lift-off phase with the thrust then vectored in a rearward direction for horizontal flight.

The program was initially funded by Canada and then later taken over by the US Air Force in 1954. The concept was evaluated, but technical problems caused it to be canceled in 1961.

The Vectored Thrust technique could also find itself used for the next-generation American fighter, the so-called Joint Strike Fighter, or JSF. Although the final selection of the prime contractor, Boeing or Lockheed-Martin, had not been made at press time, the X-32 Boeing concept will use a Vectored Thrust technique. Since it isn't a pure VTOL system, the X-32 is addressed by Chapter 14, which addresses Almost-VTOL systems.

The X-32 will use a single jet engine with two directional exhaust nozzles located at the CG of the plane for vertical or short take-offs and transition to normal horizontal flight. The third JSF entry, the losing McDonnell Douglas concept, also used the Vectored Thrust concept, which had a vectoring nozzle arrangement, and also, the complete lack of a tail. Parts of the craft bore a similarity to the F-15 Eagle. (See Chapter 13).

Vectored Thrust VTOL vehicle technology continues to be investigated in the late 1990s with a VTOL concept (the DP-2) proposed by the Office of Naval Research and DuPont Aerospace Company. Looking amazingly like the X-14 in the forward portion of the fuselage, the craft could be carrier-based and replace the E-2C, S-3, and ES-3 support aircraft.

Vertical lift would be attained from two PW530A turbofans thrusting through vectoring cascades. Then, during the horizontal flight phase, the cascades retract, creating a straight-rearward thrust stream. It was stated that up to ten tons of high density payload could be carried by such a vehicle. Following are the programs that used the Vectored Thrust VTOL principle:

COLLINS AERODYNE VECTORED THRUST AIRCRAFT (U.S.A.)

System: Collins Aerodyne
Manufacturer: Collins Radio Company
Sponsor: Company program
Operating Period: 1950s
Mission: To develop the technology to support a Vectored Thrust wingless VTOL aircraft
The Story:
During the 1950s, a number of Vectored Thrust VTOL concepts were investigated. Included in the research were programs by the Piasecki Aircraft Corporation (with its Ring Wing concept), the Avro Aircraft Ltd, Canada (with its Avrocar program), the Goodyear Aircraft Company Convoplane, and the Chance Vought Aircraft Company ADAM configuration.

One of the most interesting was the Collins Aerodyne, which was a wingless deflected thrust vehicle. The concept carried a fuselage-contained engine system with the thrust being exhausted outward on the lower portion of the fuselage. Contained in the thrust stream was a series of vanes which had the capability to turn the thrust stream downward, providing the downward lifting thrust.

With no additional onboard propulsion system, the main propulsion system served the dual capacity of both lifting and

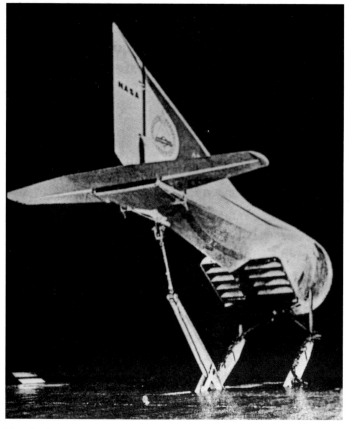

The Collins Aerodyne being tested at the NASA Ames 40-by-80 foot Wind Tunnel. (NASA Photo)

Thrust deflectors on X-14. (Air Force Museum Photo)

horizontal propulsion. Although there is no record of the Aerodyne ever making an actual flight test, the plane was tested in the NASA Ames 40x80 foot wind tunnel.

BELL X-14 VECTORED THRUST VTOL (U.S.A.)

System: Built-from-scratch prototype with significant pieces of Beech T-34 and Bonanza aircraft used in its construction
Manufacturer: Bell Aerospace Company
Sponsor: Air Force/NASA
Operating Period: 1955-to-1981
Mission: Produce a small Vectored Thrust VTOL demonstrating the principle in a successful manner.
The Story:
The story of this unique craft began in July 1955, when Bell Aerospace won an Air Force contract to construct the prototype for the Vectored Thrust X-14 program.

In an era that was seeing the design and development of sleek swept-wing fighters and bombers, the X-14 appeared to be a step backward in technology, certainly not a design possessing the look of speed or performance.

The machine appeared flimsy at best, with a spindly tricycle landing gear holding up the 25-foot long fuselage. The fixed straight wings had a span of only 34 feet, with the single vertical rear tail having a maximum height of eight feet.

The forward portion of the fuselage was totally encompassed with a pair of 1,750 pound thrust Armstrong Sidney ASV.8 turbojet engines. Later, those initial powerplants were replaced by more powerful General Electric J85 engines under a contract with NASA for their evaluation in a VTOL environment.

The thrust from those engines was routed through a movable vane system located directly below the cockpit, which amazingly was an open-style, giving the craft even more of a 1930s look about it. Obviously, such a cockpit didn't carry an ejection seat, which relegated flight testing to very low altitudes.

Fuel for those engines came from a pair of pylon-mounted external fuel tanks which were mounted up close to the fuse-

X-14 in a test rig. (Bell Photo)

The fuselage of the X-14 definitely gave the impression of this being a research vehicle. (Bell Photo)

lage. The concept, according to the company, provided ease of maintenance.

The way the engines were mounted to the forward portion of the fuselage looked as though they were added as an afterthought, since they were not faired in with the lines of the front fuselage. The X-14 resided on a spindly tricycle landing gear which hardly looked capable of supporting the craft's minimal weight.

The remainder of the strange little aircraft was a montage of parts and pieces from other aircraft. For example, a Beech T-34 Mentor provided the tail surfaces and the rear fuselage. The wing was donated from another Beech product, the Bonanza. The lightness of the initial craft, at only 3,100 pounds, was amazing. The plane, even with its small weight, was considerably smaller than an earlier VTOL researched by the company.

The craft first took to the air in February 1957 with Bell test pilot David Howe at the controls. The hovering capabilities of the plane were demonstrated. It would be over a year later, May 1958, when the first transition to horizontal flight

occurred where it also demonstrated a maximum speed of 257 knots. It would be the first VTOL aircraft to achieve vertical flight using the Vectored Thrust technique.

A strange phenomena known as "suckdown," when an area of low pressure near the ground is created during vertical liftoff, was experienced during the X-14 flight test program. The phenomena had the negative effect of tending to pull the plane back down to the ground. The solution to the problem came in an interesting way, with the lengthening of the landing gear struts.

The actual Thrust Vectoring technique was accomplished by thrust diverter vanes positioned in the engine tailpipes which exhausted under the wing. For vertical or hovering flight, the vanes were directed straight down. With this arrangement, the X-14 was then able to lift itself vertically while the plane was maintained in a horizontal attitude.

The transition to horizontal flight was a touchy undertaking which was accomplished slowly, and carefully! During this process, the vanes were brought to a rearward-pointing position very slowly and carefully until enough velocity was at-

Note the spindly landing gear on the X-14. (Bell Photo)

X-14 in flight. (Bell Photo)

The engine inlets on the X-14 were not flaired in to give the front of the craft a clean look. (Bell Photo)

With its open cockpit, the X-14 had the design look of an earlier time. (Bell Photo)

tained such that the aerodynamics of the wing could support the aircraft. Then, the vanes were turned to the full rearward position.

During hovering flight, there was a certain amount of instability that had to be addressed. In order to alleviate this situation, small air jets on each wing tip worked together to keep the wings level.

The process for reversing the process was *not* a backwards duplication of the getting-into-the-air process. First, the engines were throttled back and the engine exhaust vanes were canted downward. The aircraft then slowed until the condition existed where the vertical lift provided by the thrusting engines supported the aircraft, replacing the support of the wing, which was appreciably decreased due to the forward velocity decrease.

Interestingly, the X-14 was a USAF program for only four years before it was reverted to Ames Research Center of NASA in 1959 to address a different function. NASA would convert the X-14 into a variable stability aircraft which could artificially simulate other aircraft flight characteristics.

Later in the program, the X-14 was fitted with the aforementioned J-85 engines, which caused it to be renamed the X-14A. Then, in 1971, a more powerful version of the J-85 became available, thus another new name, the X-14B. The thrust of the modified vehicle was now 6,000 pounds, considerably greater than its previous 3,500 pounds.

Whereas the initial X-14 VTOL had a 1.1-1 thrust-to-weight ratio, the X-14B had billowed to 1.4-1, greatly aiding its flight capabilities.

The X-14 flight test program would end in May 1981 when a hard landing caused structural damage and a small fire. That was the end, after well over two decades, and the acquisition of a vast amount of data on Vectored Thrust flying from this unique VTOL machine.

The X-14 remains at NASA Ames today.

MAC-DAC/BAe AV-8 HARRIER VECTORED THRUST VTOL (U.K./U.S.A.)

System: Follow-on system of the Krestrel VTOL fighter
Manufacturer: McDonnell Douglas/British Aerospace (Hawker)
Sponsor: RAF, USMC
Operating Period: 1966-to-present
Mission: Develop a Vectored Thrust VTOL fighter using the Krestrel fighter as a baseline starting point.
The Story:
During the 1950s, there were British planners that were looking at the vulnerability of fixed military installations. They reasoned that a moving target is more difficult to hit. That brought about the need to develop a high-performance fighter capable of operating at diverse locations which did not possess fixed runways.

Note the spindly landing gears on the X-14 as it accomplishes a VTOL take-off. (Bell Photo)

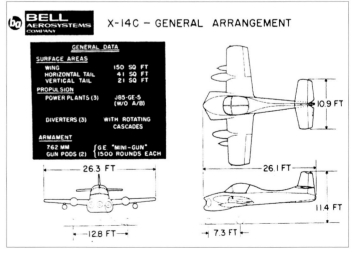

Three-view drawing of the X-14. (Bell Photo)

The XV-6A lifts off vertically from the deck of the USS Raleigh in 1966. (US Navy Photo)

The XV-6A that is currently on display at the Air Force Museum. (Air Force Museum Photo)

But the situation would be first addressed by a combination of French and American effort with the development of the previously-discussed Hawker P.1127 Krestrel program, which demonstrated a Vectored Thrust VTOL fighter with a first flight in 1960.

But it would be the Royal Air Force that would accomplish much of the work to bring the Harrier to fruition. It was decided that the P.1127 would serve as the basis for the new fighter, and to that end, nine of the P.1127s were evaluated by a unit of U.S., German, and RAF pilots.

During the same time period, a more-powerful Vectored Thrust model, called the P.1154, was canceled. A number of the P.1154 systems were incorporated into extensively modified P.1127s and were given the Harrier GR.Mk1 designation.

The new model was introduced to operational service in early 1969 with the RAF, which formed the Harrier Conver-

sion Group. Later in that year, the first operational squadron, No 1. Squadron, was formed. It was adopted by the USMC as the AV-8A, with 102 being acquired in the early 1970s.

1972 studies then indicated the need for a Sea Harrier version with the decision to proceed with full development taking place in 1975. The first Sea Harrier would fly three years later. Initial versions of the new version went to the British Royal Navy and the Indian Navy. The version was powered by a 21,500 pound Rolls Royce Pegasus 104 Vectored Thrust turbofan, and had the capability for carrying a number of different air-to-air missiles.

The successful Sea Harrier was deemed worth accomplishing a mid-life update, which included the addition of an improved Doppler radar, the capability to carry the AMRAAM missile, and a lengthened rear fuselage insert of 14 inches. With these updates, the improved Sea Harrier acquired the new name of FRS Mk2. The Mk2 had a maximum forward

XV-6A atop a test stand. (Air Force Museum Photo)

The AV-8 about to take-off on a carrier deck in 1971. (US Navy Photo)

The first British-made USMC Harrier makes a vertical landing at the completion of its test trials in 1971. (US Navy Photo)

A flight deck technician signals the pilot of a Marine AV-8A Harrier in May 1972. (US Navy Photo)

speed of 720 miles per hour and a maximum loaded weight of 26,500 pounds.

Looking at the Harrier configuration, it would be easy to assume a great similarity to the Krestrel. But there were radical differences between the two models, the Harrier being almost a complete redesign with a nav-attack system, provisions for heavier military loads, and an uprated engine.

A 1972 launch of an AV-8A Harrier from the flight deck of the USS Guam. (US Navy Photo)

The structure was generally conventional, with some use of honeycomb in the thinner parts of the airfoil surfaces and a moderate amount of titanium for weight saving. The only unconventional features were the severely dropped one-piece wing, which was removed for engine replacement, and the "zero-track tricycle" landing gear. The wing tips were bolted on, and could be replaced by longer-span panels for ferry missions.

The undercarriage was designed for a 12 feet per second landing at 16,000 pounds. Low pressure tires were used to facilitate operation from soft surfaces. The undercarriage doors were arranged to close when the gear was down in order to keep mud and debris out of the wheel wells.

The cockpit differed from a normal fighter primarily in having an improved downward field of view to facilitate approach and landing. The only additional control associated with V/STOL capability was a lever mounted alongside the throttle, to control the orientation of the four engine nozzles.

With this lever forward, the jets are directed horizontally for conventional flight. As the lever is brought back, the nozzles turn to the vertical to provide jet lift, and the nozzles can be taken slightly beyond the vertical for improved deceleration in a landing transition.

The plane's aerodynamic flying controls were comprised of a slab tailplane and ailerons powered by duplicated jacks fed from independent hydraulic systems, and an unpowered rudder. For low-speed control, the aircraft used bleed air taken from the engine to valves in the nose, tail, and wingtips, the whole system being brought into operation when the engine nozzles were rotated downwards. The opening of the reaction valves was controlled by the movement of the aerodynamic flying controls, so that the control column and rudder bar continued to operate in the normal sense. There was no sudden change in control characteristics, no complicated box

An AV-8A Harrier in flight following a 1972 launch. (US Navy Photo)

offensive capabilities of the model have increased dramatically, making it a formidable fighting machine. Carrying numerous weapon pylons under its wings, the Harrier II carried weaponry like 25mm cannons, Paveway II Smart Bombs, AGM-62 Walleyes, AGM-65 Mavericks, ASMs, bombs, dispensers, and rocket launchers.

The AV-8B was designed specifically to meet the requirements of the USMC ground commander with an aircraft based as close to the action as possible. The first operational USMC squadron was VMA-331, which was commissioned in January 1985. Full operational status was achieved the following year. In 1987, the Spanish Air Force accepted a number of the model. Other users of the model include Italy, India, and Thailand.

Some 86 British AV-8Bs operated during the Falklands War of 1982, the first time the model had seen the fire of conflict. Nine years later, USMC AV-8Bs would see action in Operation Desert Storm.

Production of the AV-8B ended in FY1994. However, during the 1990s, Boeing continued to remanufacture standard AV-8B and AV-8B(NA) models to the advanced AV-8B(PLUS)

of tracks for mixing the control inputs, and no gyroscopic precession effects.

In another phase of the longstanding Harrier story, during the mid-1970s, there was a joint effort between British Aerospace and McDonnell Douglas to develop an Advanced Harrier. But the joint effort was short lived, with the British withdrawing in 1973, causing McDonnell Douglas to go it alone.

Then, in 1981, the Royal Air Force accepted the AV-8B Harrier II design that McDonnell Douglas had continued to develop. The program, though, had also faced money problems as the company had been concentrating its efforts on its new F/A-18 fighter and had actually laid dormant for three years before being revived.

A number of changes had been made to the AV-8B with a more powerful 22,000 Pratt & Whitney F402 Vectored Thrust turbofan with square-cut front nozzles which tend to increase lift in the VTOL mode. The inlets were also redesigned for increased thrust. The new airframe incorporated a new longer-span wing with a supercritical section and less sweep. With a goal of increasing the range, the internal fuel capacity was increased by an amazing 50 percent. The actual lift of the model was considerably increased with new lift-improvement devices under the fuselage, along with large flaps that were lowered during the vertical lift phase. The devices also allowed for improved performance during rolling take-offs.

Even though the Harrier has always been considered a bit of an oddity against its normal-type fighter brothers, the

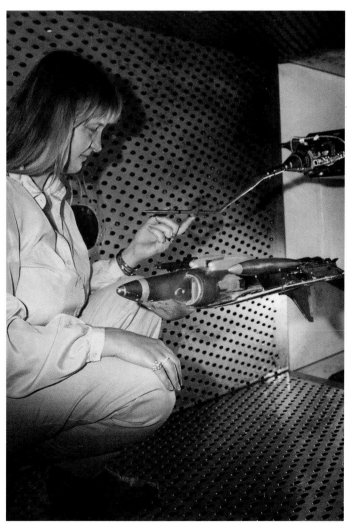

Testing in 1981 examined the effects of ordnance separation from the Harrier aircraft. (USAF Photo)

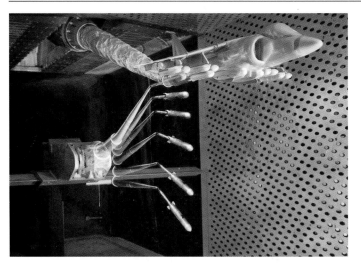

A wind tunnel model of the AV-8A is installed in an Air Force wind tunnel. The 1996 testing involved the AMRAAM missile.

A drawing showing the evolution of the AV-16A. (McDonnell-Douglas Photo)

configuration. The Marines also acquired a number of TAV-8B trainer versions.

GLOSTER METEOR IV VECTORED THRUST VTOL (U.K.)

System: Based on Meteor IV fighter
Manufacturer: Gloster, Westland
Sponsor: Royal Air Force
Operating Period: 1952-to-1957
Mission: Served as a prototype for testing and proving a Vectored Thrust fighter concept.
The Story:

Its serial number was RA490, a Gloster Meteor IV twin jet British fighter that had served in operational service with the RAF from 1948 until 1952. The model was equipped with Vickers F2/4 Beryl engines. Following this service, RA490 was directed into a research where it was modified to investigate the effectiveness of a Vectored Thrust system.

The Westland Aeroplane Company was designated to accomplish the modification. First, new Nene 101 engines were substituted for the Vickers engines, greatly increasing the craft's thrust capability. In order to accommodate the longer powerplants, the engine nacelles were lengthened. The configuration was interesting in that the new engine mounting placed the engines far ahead of the plane's center of gravity, giving a perception of possible aircraft instability.

The rear of the engines were equipped with so-called "jet pipes" which stretched through the rear of the engine nacelles. Contained within the pipes were movable valves which could change the direction of the thrust, either straight out the back or straight down.

Testing, which was accomplished at Farnborough in the mid-1950s, showed that it was possible to fly the vehicle without stalling at about 30 knots slower speed than its original configuration.

The revolutionary concept introduced by this program, though, would never reach fruition. In fact, the plane closed its career in a non-research capacity, actually as a water bomber practice aircraft.

An artist's concept of the AV-16A. (McDonnell-Douglas Photo)

The Gloster Meteor IV RA 490 was fitted with a Deflected Thrust system to reduce landing speed. (Gloster Photo)

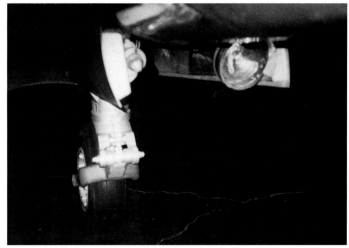

The leading edge of the P.1127 wing joins the fuselage of the aircraft. (Bill Holder Photo)

A view of the P.1127 front landing gear. (Holder Photo)

HAWKER P.1127/V-6 KESTREL VECTORED THRUST VTOL (U.K.)

System: Built-from-scratch prototype
Manufacturer: Hawker Aircraft Ltd., Bristol
Sponsor: RAF, NASA
Operating Period: 1957-1968
Mission: Develop a small VTOL fighter using the Vectored Thrust technique
The Story:
The P.1127 Kestrel Vectored Thrust fighter program began in 1957 with Hawker and Bristol being the prime contractors in the effort. The huge United States interest in the program saw significant U.S. money contributed to the program.

The program was probably unofficially initiated by the studies of noted French aeronautical engineer Michel Wibault, who proposed a concept for a flat-riser Vectored Thrust fighter aircraft. Unable to generate any interest in French circles, Wibault turned to the United States, having worked for Republic Aviation during World War II, and got interest from USAF officials, who in turn consulted with Bristol Aero Engines of the U.K.

The Bristol Pegasus engine was selected as the propulsion system for the new fighter. An interesting design, the engine has a so-called "bifurcated jetpipe" with vectored thrust front and rear nozzles. The engine also featured a common intake, counter-rotating spools, and cascade-type louvers in each of the four nozzles. During the early years of the program, the engine was refined a number of times and finally evolved to the BE52/2 version, which was the first version to take to the air.

Lightness and strength were two high priorities in the craft's structure, with the expected outriggers located on each wingtip. That wing, by the way, bore a marked resemblance to that used on the US Navy A-4 light attack aircraft. The landing gear was a tricycle style, which provided the plane with a nose-high attitude.

Every pound saved on this design was important in this program. As one engineer stated, "That one extra pound on the plane could be the pound that keeps the plane on the ground." To that end, the wing structure finally stabilized at an impressive six pounds per square foot of wing area. When you figure it out, that weight is lighter than if the wing had been fabricated of super-light balsa wood!

With the thrust directed down along the centerline of the aircraft, typical with the Vectored Thrust technique, additional control to maintain stabilization was required. In the case of this program, it was achieved by a reaction control system, in this case, using a constant total bleed system. Small reaction thrusters, two on the wingtips and one each on the nose and the tail, provided this control.

The prototype P.1127. (Hawker Photo)

A close look at the pair of deflecting thrust nozzles on the P.1127. (Bill Holder Photo)

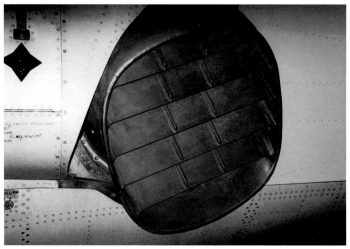

Note the weld marks on the thrust deflectors. (Bill Holder Photo)

That important reaction system would be refined later in the program to a "fully variable bleed" system and would operate only in response to commands from the pilot. Differences in this RCS system included a separate yaw thruster, higher stick gearing, and increased control powers, all pointing toward greater pilot control.

The first untethered flight test of the P.1127 took place in 1960, but the results were not considered adequate for evaluating actual aircraft control response. It was for that reason that an untethered flight was made shortly thereafter, followed by the first double transition, occurring in September 1961.

Problems during this early testing revolved around the lack of control around all axes, which, when associated with limited height control power, caused a high pilot workload while the craft was in the hover mode.

As with all aircraft of this type, the additional consideration of ingesting the hot engine exhaust gases had to be considered, but there was an interesting method of overcoming that problem. Taking off and landing at a slight forward speed eliminated that problem.

Later in the program, the vertical stabilizer was converted to a swept configuration, a change that would be adopted for the Kestrel in the 1963-1964 time period.

Propulsion improvements included an improved Pegasus 3 powerplant, which generated 13,500 pounds of thrust, to be later followed with the Pegasus 5 engine with 18,000 pounds of thrust, the performance of the plane being greatly increased. The addition actually provided Mach 1+ horizontal flight capabilities.

Later, the U.K., U.S., and Germany initiated a program to fund nine improved P.1127s for use by a tri-nation squadron where operational trials were conducted.

In the late 1960s, one of those planes was assigned to the Ministry of Defense, which was using it to evaluate an even more-advanced version of the Bristol Siddeley Pegasus engine. But things did not all go well, as the plane crashed during a ferry flight after an engine flameout. Fortunately, the pilot was able to safely eject.

In the United States, the National Aeronautics and Space Administration (NASA) obtained two P.1127s for flight testing

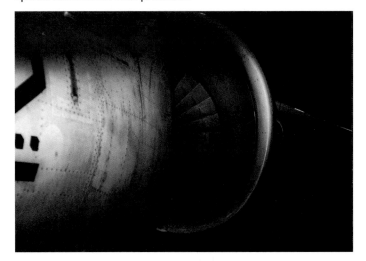

A head-on look into the intake of the fuselage-mounted engine intake. (Bill Holder Photo)

A front quarter view of the Yak-36. This airplane is on display at the Russian Air Force Museum outside of Moscow. (Steve Markman Photo)

at Langley Research Center. During this activity, the planes were given an XV-6A designation.

One of those Kestrels crashed in August 1967 with its fuselage and wings being heavily damaged, making repair impractical. It was decided that the reusable parts of that plane would be used as spares for the remaining plane, which was flown in a research mode.

In retrospect, probably the most important accomplishment in the Kestrel program was to lay the background and establish the technology for the family of Harrier fighters that would follow.

YAK-36 FREEHAND VECTORED THRUST VTOL (SOVIET UNION)

System: Built-from-scratch prototype
Manufacturer: Yakovlev
Sponsor: Soviet Air Force(SAF)
Operating Period: 1962-1971
Mission: Development of a twin-engine fighter capable of vertical take-off and landing capabilities using a vectored thrust technique.

The Story:
Following the start of the American X-14 program, the Soviet Union attempted the development of a similar concept, but the technique was employed in a conventional-looking fighter configuration.

The plane's twin Tumanskiy/Khatchaturov 11,000 pound thrust R27 turbojet engines were mounted low in the forward fuselage, requiring the front fuselage to be extremely wide and giving the design a bit of an awkward look. Also, the cockpit was of a standard type, but located forward of the plane's center of gravity.

The VTOL capability of the craft was obtained by rotatable engine nozzles that were located below the plane's center of gravity and a surprising four feet apart. The nozzles were equipped with louvers which could be vectored through the required 90 degrees, with the exhaust passing through the craft's center of gravity. At lift-off, the plane had a thrust-to-weight ratio of 1.06-1. Testing showed problems with the engine inlets ingesting the hot exhaust gases, so a large retractable apron was placed under the engine inlets.

As is typical of the vectored thrust concept, separate propulsion was required for pitch and yaw control. For the Freehand, that control was acquired by a series of small reaction jets, one each located on the wingtips, a third in the surprising location of the tip of the refueling nose boom, and a fourth on the tailcone. The thrust for these units was provided by bleed air from the engine.

The overall length of the fuselage was 57 feet, six inches, with a height of 14 feet, nine inches. The wing span was only 27 feet with a 40-degree sweep. Weighing over 12,300 pounds empty, the maximum weight grew to 20,700 pounds. With a maximum range of 200 nautical miles, the Freehand had a top speed of about 540 knots, both at sea level and altitude. The ceiling was a reported 39,375 feet.

During the development program, four prototypes were constructed, one ground test, two flight test, and one static test. From those numbers it would appear that the designers had extreme confidence in this aircraft with a full testing of its capabilities undertaken.

The fact that the plane was jet-powered was unique from designer Yakovlev's background as most of his background was with the design and development of prop-powered fighters, a number of which achieved success during World War II.

Aft view of the Yakovlev Freehand VTOL subsonic fighter. (Paul Nann Photo)

Testing for the new plane began in 1962 with tethered tests conducted over a metal grid covered pit. Initially, the plane was equipped with smaller R11 engines, but it soon was realized that larger engines were necessary, hence the substitution of the R27 engines.

The first untethered flight took place in January 1963, with the first double-transition occurring in September of the same year. The program suffered a setback in 1964 when the program's chief test pilot was injured in another test program, requiring a new pilot for substitution.

As the program progressed, problems were noted when the craft was within five feet of the ground—it had a tendency to oscillate. In 1966, one of the flight test versions made a hard landing, breaking an outrigger.

During the test program, changes were made to the bleed outlets on the wing pitch/yaw nozzles. Also, plenums were added in the wingtips to ensure an even flow of the bleed air.

It wouldn't be until 1967, at the Domodedovo Air Show, that the Yak-36 would be exposed to the world. Interestingly, the version that was shown at Domodedovo carried rocket pods, the exact reason not being revealed.

Following the testing of the initial Yak-36 version, there was also a reported investigation of a Navy version which was called the Yak-36M, supposedly for "Modified."

Reportedly, the Yak-36 was able only to do vertical take-offs and landings, with no horizontal flight capability. In restrospect, the Yak-36 served as a technology demonstrator which would eventually lead to the Yak-38 Forger, which would be an operational fighter.

6
Lift-Cruise VTOL Systems

Lift-Cruise VTOL aircraft, sometimes called "dual propulsion" VTOL aircraft, had dedicated engines that were used for lift only. A cruise engine or engines provided for forward flight and also could double as a lift engine. Of course, nothing in this world is completely pure…on some of these aircraft, the lift engines could provide a small amount of forward thrust. There are two major sub-categories: Lift + Cruise and Lift + Lift Cruise.

Aircraft in the Lift + Cruise category used separate engines for hover and for cruise flight. They included only the Dassault Balzak and Mirage III-V, and the Short Brothers S.C.1. These aircraft used the lift engines only for hover and transition, and the cruise engine only for forward flight. The cruise engine did not assist during hover.

The Mirage III-V was to be a VTOL strike fighter for the French Air Force, but it never progressed beyond two prototypes being built and tested. The Balzak, which looked very much like the III-V, was a technology demonstrator aircraft built from the prototype Mirage III. It incorporated many of the features planned for the III-V, in order to determine if they would work and then to mature the technology. The Short Brothers S.C.1 was built strictly as a research aircraft with the mission of investigating Lift + Cruise technology and solving its associated problems.

The Lift + Cruise concept provided for a relative level of simplicity in that separate engines provided for hovering and cruise flight. This in turn allowed for cruise engine inlets and exhaust nozzles to be optimized to produce maximum power during forward flight, and eliminated the weight and complexity of thrust diverters.

Of course, there were negative features, also. With the lift engines used only for vertical take-off and landing, they were dead weight throughout most of the mission. They also occupied fuselage volume.

The Lift + Lift Cruise sub-category improved somewhat on the Lift + Cruise sub-category in that the cruise engines doubled as lift engines. Six aircraft fall into this one concept: the Lockheed XV-4B, VFW VAK 191B, Dornier Do 31, EWR VJ 101C, Yakovlev Yak 38, and Yak 141.

By using the cruise engine to obtain vertical thrust for hover and transition, the amount of thrust that the lift engines needed to provide can be reduced. This gives designers added flexibility in choosing the number and size of the lift engines.

Two different methods were used to allow the cruise engine to produce both lift and cruise power: use of a diverter valve that directed the thrust either straight back or out the bottom, or use of vectoring exhaust nozzles that turned to direct the exhaust in the desired direction. Some authors chose to categorize the later under the Vectored Thrust concept. The authors felt that the defining element was the use of two different propulsion systems and the means of obtaining the dual function from the cruise engine was of lesser importance. The important distinction in this concept is that the cruise engine performed double duty and worked in conjunction with one or more lift engines. In some cases, the lift engines produced most of the hover thrust and were assisted by the cruise engine. In other cases, the cruise engine provided most of the lift thrust, and one or two lift engines provided the needed additional vertical lift.

DASSAULT MIRAGE III-V AND BALZAC LIFT + CRUISE VTOLS (FRANCE)
System: Based on Mirage III-E
Manufacturer: Dassault
Sponsor: French Air Force
Time Period: 1960 - 1967
Mission: Nuclear strike fighter

Looking forward on a Dassault Mirage III-V.

The Story:

With the British developing the P.1127 as a technology development aircraft for what eventually became the Harrier, France jumped in with an early commitment to develop a VTOL strike fighter for the French Air Force. In 1960, they committed to develop a VTOL version of the Mirage III, designated the Mirage III-V. The main mission would be all weather, day/night, reconnaissance, and tactical nuclear attack at low altitude and high speed. At the same time, they committed to build and test a technology demonstrator aircraft, designated the Balzac, to reduce development risk. The French Air Force ordered four prototypes, but only two, plus the demonstrator, were built and flown.

The Mirage III-V's general layout was very close to the Mirage III-E. The concept selected to obtain vertical flight was to use Lift + Cruise engines. The single seat fuselage was longer than the III-E's to accommodate four pairs of lift engines. The low, delta wing was of all metal construction and utilized stressed skin machined panels with integral stiffeners. While very similar to the III-E's wing, the III-V's differed by having a break in the leading edge, giving it a compound sweep. Overall dimensions were a wing span of 28.6 feet and a length of 59 feet. The maximum take-off weight was 29,630 pounds.

All landing gears had twin wheels, whereas most other Mirage models up to that point had a single wheel on each landing gear. A grille intake panel hinged at the rear covered each pair of lift engines. Attitude control during hover utilized compressor bleed air directed through dual nozzles in the nose, tail, and wing tips. The only external difference between the two prototypes was that the second one had sideways-opening air intake doors over the lift jets. Planned armament was to include nuclear weapons, with the provision for a single large store recessed into the bottom of the center portion of the fuselage.

Main propulsion was to be provided by a SNECMA TF-106 afterburning turbofan, a derivative of the Pratt & Whitney JTF10, producing 20,500 pounds thrust, but it never was installed in either of the prototypes. The first prototype used a SNECMA TF-104, also a derivative of the JTF10, producing 13,800 pounds thrust, because the TF-106 was not available. The second prototype used a Pratt & Whitney TF-30 turbo-

Note the open doors on the upper fuselage of this Mirage III. (Dassault Photo)

fan engine rated at 11,330 pounds and 18,520 pounds with afterburning. Vertical propulsion was by eight Rolls-Royce RB.162-1 lift engines producing 3,525 pounds static thrust each, mounted in pairs in the fuselage. The third and forth prototypes, which never were built, would have had a SNECMA TF306 turbofan engine producing 19,800 pounds thrust, and four Rolls-Royce RB.189 lift engines.

The first prototype began hovering trials on February 12, 1965, at the French flight test center at Melun-Villaroche. It made the first transition from horizontal to hovering flight on March 24, 1966. Eventually, it achieved a speed of Mach 1.35 at high altitude with the TF-104 engine. The second prototype made its first flight on June 22, 1966, but was destroyed in an accident on November 28, 1966.

By 1967, the French Air Force decided not to build the remaining two prototypes originally planned nor to proceed with production.

The Balzac was a technology demonstrator used to reduce the development risk for the Mirage III-V. Its primary role was to develop the Lift+Cruise concept by investigating stability at low speed, hover, and transition, and to test various components and concepts that would go in the Mirage III-V. The Balzac was built from the prototype Mirage III.

The configuration obviously was very similar to the Mirage III-V, but the gross weight was only 14,500 pounds. Initially, the Balzac had a non-retracting landing gear, but subsequently it was replaced with one that retracted. The main

Mirage III-V 02 fighter. (Dassault Photo)

Note the shorter length fuselage, the Dassault Balzac having been built from the original Mirage IIIV prototype. (Dassault Photo)

gear had dual wheels, but the nose gear had a single wheel. Compressor bleed air directed through dual nozzles in the nose, tail, and wing tips provided attitude control, as in the III-V. Other features included a Martin-Baker AM-6 zero speed rocket ejection seat and a single throttle that controlled all eight lift engines. The lift throttle was positioned so that the pilot could operate both the lift and cruise throttles simultaneously with the same hand.

A Bristol Siddeley Orpheus turbojet engine of 4,850 pounds static thrust provided cruise power. Eight vertically-mounted Rolls-Royce RB.108 lift engines of 2,200 pounds static thrust each (2,000 pounds net each, after allowing for compressor bleed air needed for hovering control) provided vertical thrust.

The Balzac made its first tethered flight on October 12, 1962, and its first free vertical flight on October 18. In November, it performed a demonstration for the press at the French flight test center. The first transition occurred on March 18, 1963. The Balzac crashed during a flight test on January 10, 1964, but was repaired and resumed flight testing within a year.

SHORT BROTHERS S.C.1 LIFT + CRUISE VTOL (UNITED KINGDOM)

System: Built-from-scratch research aircraft
Manufacturer: Short Brothers & Harland
Sponsor: Ministry of Supply
Time Period: 1954 - 1971
Mission: Investigate Lift Cruise technology issues
The Story:
In one of the earliest research efforts in VTOL flight, Britain's Ministry of Supply awarded a contract for the design and construction of two S.C.1 aircraft to Short Brothers & Harland, of Belfast, North Ireland, in August 1954. The specification, ER.143, was relatively short. It called for a research aircraft capable of making vertical take-offs using jet lift only and then accelerating into conventional flight supported by the wings alone without use of the lift engines. It also specified the ability to decelerate to a stationary hover and make a vertical descent to landing using the jet lift engines. Given that there was no precedent for a vehicle with such a flight envelope, this specification was very bold for its time.

Shorts' design carried the internal designation of P.D.11. The design philosophy, and the bulk of the design effort, was to build an aircraft that could operate vertically and perform the transition to and from conventional flight. Requirements for conventional flight received much less emphasis. Light weight, to give the best ratio of thrust to weight, was a primary requirement. Essentially, the aircraft had to have the smallest size possible to carry a pilot, five engines, instrumentation, and fuel.

The S.C. 1 was the first fixed-wing VTOL aircraft program in the United Kingdom. Two aircraft were built, and both survived and eventually were retired to museums. They carried the tail numbers XG900 and XG905.

The S.C.1 was a rather chunky looking delta wing aircraft, with a wing span of 23.5 feet, length of 24.4 feet, height of 9.9 feet, and wing area of 211.5 square feet. It was flown by a single pilot. Power was provided by five Rolls-Royce RB.108 turbojet engines. Four were mounted vertically in the fuselage for lift and one in the tail for cruise. All engines were identical. The cruise engine's air inlet was located at the base of the dorsal fin. Separate ailerons and elevators were located on the trailing edge of the wing, and did not double as flaps. The lack of internal space within the airframe was always a problem, especially as upgrades were made throughout the S.C.1s' lives. As a research aircraft, it was designed purely for low speed flight. The maximum demonstrated speed at sea level was 246 miles per hour, and the sea level rate of climb was only 700 feet per minute.

The stubby fuselage was of all-metal, semi-monocoque structure. It housed the four lift engines just aft of the cockpit.

Short S.C.1 on display at Fleet Air Arm Museum in England. (Paul Nann Photo)

A wire mesh screen over the engine intake bay on top of the fuselage protected the lift engines from ingesting foreign debris. Below the wire mesh were longitudinal louver doors that covered the lift engines and faired in the fuselage for conventional flight. They were spring loaded to the closed position and were pulled open by the air flowing into the lift engines. Titanium sheet surrounded the inside of the engine bay to increase fire resistance. The lift engines' exhaust nozzles protruded slightly below the bottom of the fuselage. Fairings were positioned around the leading edges of each nozzle to streamline the air flow during forward flight. A five foot long boom in front of the nose carried the pitot head, pitch, and yaw vanes. The delta wing had a thickness of 10 percent of the chord and the leading edge was swept 54 degrees. Fuel was carried in the wing leading edges and in bladder tanks between the wing main spars. Total fuel capacity was 264 gallons. The maximum gross weight was 8,050 pounds, but a lower weight of 7,800 pounds was typical for vertical take-offs. The maximum landing weight was 6,800 pounds.

The spacious, unpressurized cockpit was enclosed by a large helicopter-like canopy, giving excellent visibility, including downward, and in all directions except for straight backward. The instrument panel was positioned low in the cockpit to further improve the forward visibility. A head-up display was mounted to the instrument panel to provide essential flight data to the pilot while he kept his vision focused outside the cockpit. A Folland lightweight ejection seat originally was installed, but was replaced with a Martin-Baker Mk V4 seat with zero-zero capability. Flight controls consisted of conventional control stick and rudder pedals. All four lift engines were controlled by a single lever resembling a helicopter collective pitch lever.

Both S.C.1s were powered by five Rolls-Royce RB.108 direct lift engines. The RB.108 was the world's first direct lift engine, designed specifically to operate in the vertical position. It was lightweight, with an 8:1 thrust/weight ratio. Initially, it produced 2,010 pounds of static thrust at sea level, but the power output gradually increased to 2,130 pounds over the life of the program. The four lift engines could be tilted fore and aft of vertical through a range of 35 degrees to assist transition to and from level flight.

The cruise engine had to be mounted at a 30 degree angle because the lubrication system would not function had the engine been mounted horizontally. Curved ducting directed air into the intake and ducted the exhaust out horizontally. The cruise engine was started with compressed air provided by an external source. The lift engines were started either by bleed air from the cruise engine, or, at high forward speeds, by ram air taken in through the gill intakes on the top of the fuselage just aft of the cockpit. The gill intakes were manually operated by the pilot, and normally were closed during conventional flight.

The S.C.1 had a fixed tricycle landing gear. The wheels, brakes, and tires were made by Dunlop. The drag brace on the two main gears could be adjusted manually on the ground

to position the main struts forward for normal take-off and landing or rearward for vertical landing. The different positions were required because of the risk of tipping backward onto the tail when making vertical landings. The twin wheels on each main strut could castor freely when the gear was in the rearward position for vertical take off and landing. When the gear was in the forward position for conventional take-offs and landings, a hydraulically actuated lock held the axle rigid so the aircraft could roll straight forward. There was no nose wheel steering. The nose wheel castored freely, and the S.C.1 was steered using differential wheel brakes. The struts were fitted with fairings, although these were left off many times during the lives of both aircraft. A drag chute was used to assist braking on conventional landings.

All control surfaces were moved by push/pull rods. Irreversible power units assisted in moving the surfaces. The elevator had an electrically actuated trailing edge tab for trim.

The three-axis stability augmentation system was developed by the Royal Aeronautical Establishment (RAE) at Bedford. It was powerful enough to flip the aircraft upside down in 2 seconds should a hardover occur. For this reason, the system was built in triplicate, with a voting logic such that a defect in any of the systems would cause that circuit to be removed from the control loop. RAE demanded that the pilot must be able to change rapidly from automatic to manual control and vice versa and that the system must be fail safe. No single fault in the system should be able to cause a catastrophic failure.

Bleed air from each of the lift engine compressors was used to control attitude during hover and at very low speed. Magnesium ducting ran from the engines to the nose, tail,

The Short S.C.1 in flight. (Shorts Photo)

and to the bottom of the wing, just inboard of the tips. The nose and tail jets not only controlled pitch, but could be rotated from side to side to control yaw. All four nozzles normally were partially closed. The control action consisted of differential opening and closing of opposite pairs.

First metal was cut in spring 1955, and XG900 was completed in late November 1956. Only the propulsion engine had been installed, since XG900 would first fly as a conventional aircraft to evaluate its performance and handling. Engine runs began on December 7, 1956. Taxi tests began ten days later and lasted for three months. It was decided that the company airfield at Syndenham was not adequate for the S.C.1's maiden flight and that flight testing should be performed at the Royal Air Force's flight test center at Boscombe Down. On March 6, 1957, XG900 was loaded onto a ship for the two day trip to Southampton. From there, it was taken by truck and arrived at Boscombe Down on March 11.

In the meantime, two Thrust Measurement Rigs, or "flying bedsteads," were built and used to gain experience with using thrust control for hovering. Basically a truss structure with Nene engines for lift, it used engine bleed air for reaction control. One crashed in September 1957, but was rebuilt for static display at the Fleet Air Arm Museum, Yeovilton. The second crashed two months later, killing the test pilot. Use of these devices then was stopped, since the principle of using vertical thrust and bleed air for control had been proven.

XG900 made its first conventional take off and landing on April 2, 1957. Touchdown for a conventional landing was at a noticeably high angle of attack because of the lack of a wing flap. It did not fly particularly well in the conventional mode. For this reason, testing proceeded very carefully, and the envelope was expanded in a great number of very small, discrete steps. Although the flight characteristics had many shortcomings, they were determined to be acceptable, considering that this was a research aircraft. The S.C.1 particularly demonstrated poor directional stability, which partially was cured by the addition of the dorsal fin at the base of the vertical tail. This fin was also added to XG905 while it was still being built. Following completion of conventional test flights, XG900 was returned to Belfast for installation of its lift engines. Other modifications included changes to the canopy, replacing the side entry door with a hatch on the top.

In the meantime, XG905 was completed and had its lift engines run on September 3, 1957. Tethered hover flights began on May 23, 1958, and were followed by the first free hovering flight on October 25, 1958, over a gridded metal platform located 12 inches off the ground to limit ground erosion and reingestion of debris. The first landing away from the platform was made a month later. In hover, the S.C.1 flew almost like a helicopter, the pilot using the control stick for pitch and roll and the throttle as a collective control. However, when hovering, the tires would melt after 90 seconds, surrounding air temperatures became very warm within 90 feet, and the recirculation of hot exhaust gas reduced the lift engine thrust. It was also noted that a few seconds of hover-

ing at 20 feet or less was enough to boil the tar between the slabs of concrete on the ramp.

By September 1959, XG905 had flown extensively in hover up to a maximum speed of 40 mph, and in conventional flight up to 170 knots and down to a level stall speed of 134 knots. At this point, there was enough confidence in XG905 that it was taken to the air show at Farnborough in September 1959. XG905 demonstrated hovering, but suffered an embarrassing incident. As it hovered over a field of newly mowed grass, the clippings got sucked into the recirculating flow around the aircraft. As the cloud of clippings rose and eventually got sucked into the engine screen, the engines lost power and the aircraft settled abruptly onto the grass. No damage resulted, other than a few red faces.

XG905 made the first transition from vertical to horizontal flight on April 6, 1960, at RAE Bedford. The maneuver was performed at an altitude of 100 feet. In performing the build up, the transition was approached from both ends of the speed regime; in hover, speed was increased in 10 knots increments up to 80 knots, and in conventional flight, from 140 down to 75 knots. On April 6, 1960, XG905 made the first complete transition to and from hover. The aerodynamic controls began to take effect at 30 knots, and the minimum wingborne speed was 134 knots. The lift engines could be shut down at 140 kts. To transition from cruise to hover, the propulsion engine was shut down and the lift engines moved to a forward position to slow the forward speed. This was very effective, and the lift engines were never tilted all the way forward. The S.C.1 could be brought from 140 kts to a complete stop in 1,200 feet. By September, XG905 was back at the Farnborough air show for the first full transition to be performed in public.

Getting back to XG900, the installation of its lift engines at the Short Brothers factory was not completed until sum-

Short S.C.1 in flight.

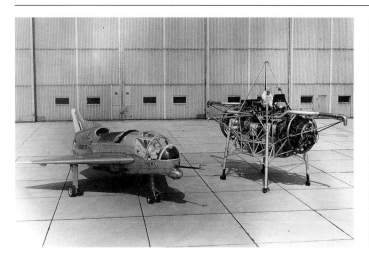
The Short S.C.1 prototype sits next to its hover test rig.

A view of one of the VJ-101's wingtip-mounted engines. (Markman Photo)

mer 1960. In April 1961, it was handed over to RAE Bedford and flew at Farnborough in September 1961. It remained at Bedford through the 60s.

Testing of XG905 continued at the Short Brothers' facility at Belfast, making over 80 flights during a three month period. Unfortunately, it was involved in a fatal crash on October 2, 1963. The cause of the crash was a vertical reference gyro failure in the stability augmentation system. The system correctly sensed the fault, but it did not properly "vote out" the failed system, despite the triple safeguards insisted on by RAE. The failure occurred at an altitude of less than 30 feet, and despite the pilot's attempt to revert to full manual control, there was not enough altitude to allow him to regain control. XG905 flew into the ground and rolled over, killing the pilot. The aircraft returned to the factory for repairs, but didn't fly again until 1966. During the rebuilding, Short updated the stability augmentation in both aircraft to prevent a similar failure from happening again.

When XG905 returned to service in 1966, it joined XG900 at RAE Bedford for all-weather and night flying trials. Both aircraft were operated by the Blind Landing Experimental Unit. After two years studying VTOL operations in bad weather and at night, XG900 was restricted to ground running only because the airframe life had been exhausted.

As it rises almost vertically, the VJ-101 seems to defy the laws of gravity. (Markman Collection)

Both aircraft remained at Bedford until their retirement in 1971. XG900 was given to the Science Museum at Hayes, Middlesex, in June 1971. It was displayed for only a few months before being placed in storage, pending a more suitable location. It eventually found a home at the Fleet Air Arm Museum at Yeovilton. XG905 was given to the Ulster Folk and Transport Museum, Belfast, in May 1971, and went on display in June 1974.

EWR VJ 101C LIFT + LIFT CRUISE VTOL (GERMANY)
System: Built-from-scratch prototype
Manufacturer: Entwicklungsring Süd (EWR)
Sponsor: German Ministry of Defense
Time Period: 1959 - 1971
Mission: Prototype VTOL interceptor
The Story:
In 1955, the ban on aircraft development and production in Germany ended. The German Air Force needed an interceptor aircraft that took into account Germany's size and geographic location as a front line state. This consideration made high top speed and excellent rate of climb primary requirements. By the autumn of 1957, the Ministry of Defense added VTOL capability as a requirement. Under government encouragement, Bölkow, Heinkel, and Messerschmitt formed a company called Entwicklungsring Süd (EWR for short, the name roughly meaning "Development Group, South") in February 1959. Their objective was to take the program from design to production. A contract was signed for an experimental interceptor aircraft without armament designated the VJ 101C (VJ being short for Vertikalstartendes Jäger, or V/STOL Fighter). Five prototypes were planned, but only three went into detailed design, and only two were built and flown. The two prototypes, X1 and X2, were to investigate hover/transition and high-speed flight, respectively. Preliminary design work started in 1959. The consortium split the work such that Messerschmitt would build the fuselage, Heinkel the wings, and Böelkow several accessories and other items. In addition, a total of

The fuselage of the VJ-101 closely resembles that of the Lockheed F-104. (Markman Collection)

115 foreign companies participated in varying amounts, including seven from England, four from France, and three from the United States.

The VJ 101C's overall configuration was for a slender, single seat aircraft with a thin, high wing having swept leading and trailing edges. Fuselage mounted lift engines and swiveling lift/cruise engines mounted on the wing tips provided vertical thrust. The fuselage, tail, and landing gear bore a general resemblance to the F-104. This was not surprising, since Heinkel was building F-104s under license. The VJ 101C's overall length was 51 feet, 6 inches, and the height was 13 feet, 6.5 inches. The thin wing, with a 5 percent thickness ratio, had an area of 200 square feet and a leading edge sweep of 39 degrees. The wingspan was 21 feet, 8.5 inches, including the engine pods.

Overall, the structure consisted of light alloy, with steel and titanium used near the engines. The nose housed telemetry equipment, although this area would house radar in a production version.

A unique "three point suspension" system that used six Rolls-Royce RB.145 engines provided the vertical lift. Two were mounted vertically just aft of the cockpit, and two each were mounted in a swiveling pod located at each wing tip. The swiveling engine concept gave the advantage of not suffering thrust loss due to the exhaust being deflected downward from a rigidly fixed engine. The three-point suspension gave other advantages. The thrust modulation replaced compressor bleed air for pitch and roll control. This reduced weight and complexity by eliminating the bleed air ducting that ran through the fuselage and wings. Also, differential swiveling of the pods through +/- 6 degrees provided yaw control. In developing the swiveling pod concept, EWR engineers made use of earlier preliminary design work by Bell on a similar, but unsuccessful VTOL fighter they designed for the United States military.

Although the VJ 101C lacked any armament, it contained systems such as built-in self-checking circuits and internal starting capability to allow operation from sparsely equipped, dispersed operating locations. Other systems included a Martin Baker Mk GA7 ejection seat with 0-0 capability, hydraulic controls by Dowty Rotol of Great Britain, hydraulic pumps and starters by Vickers of the U.S., and wheels, tires, and brakes by Dunlop of Great Britain. Three fuselage tanks held all fuel. All engines fed from the forward tank under automatic control, with manual back-up.

The X1 prototype, intended primarily to investigate vertical take-off, landing and transition, had a planned top speed of Mach 1.08 at 10,000 meters and up to Mach 1.4 in a 10–15 degree dive. Empty weigh was 9,100 pounds with a maxi-

The VJ-101 generated a lot of interest whenever it was displayed at an international air show. (Markman Collection)

mum vertical take off weight of 13,200 pounds. The X2's mission was to investigate high-speed flight, and its wing tip engines had afterburners that increased the thrust to 3,650 pounds each. The extra power improved the flight performance and take off weight, but also increased problems with recirculation of hot gasses and ground erosion. X2's empty weight was 11,990 pounds, and its maximum vertical take off weight was 17,630 pounds using afterburner. It carried additional fuel in the rear fuselage. Top speed for level flight was Mach 1.14+ at 32,800 feet. The X3 prototype was to be a two seater, but never was built.

Rolls-Royce and MAN Turbomotoren GmbH of Munich jointly developed the RB.145 engines specifically for the VJ 101C. The design derived from the RB.108 lift-jet engine and was first tested in April 1961. Without afterburner, it produced 2750 pounds of static thrust, which increased to 3,650 pounds with afterburner. The afterburning engines carried the designation RB.145R. Rolls-Royce/MAN built twenty-two RB.145 engines for the VJ 101C program, and they accumulated a total running time of 4,011 hours.

At zero airspeed and full power there was insufficient air flow into the engines. To increase the air flow, the forward portion of the engine pod slid forward to form an annular auxiliary intake to avoid choking during hover or vertical take-off.

The pods rotated about a hollow axle. The axle housed the control rods for engine control, lines for fuel, and hydraulic fluid. Duplicate hydraulic pistons rotated the pods. To eliminate the need to duct compressed air through the axle, hydraulic power was used to start the engines. Rolls-Royce designed the complete pod assembly.

Honeywell built the stability augmentation system. It used rate damping with attitude feedback. This mechanization provided stick steering during hover, which means that the VJ 101C tended to move in the direction that the pilot moved the stick, rather than simply pitching or rolling. In forward transition, the stabilization converted to only rate damping by the time the pods were rotated to 45°, which put the pilot back into controlling the pitch and roll motions directly.

A single, collective throttle lever was used to increase/decrease vertical thrust during hover. Both prototypes used differential thrust to control pitch and bank attitude during hover, eliminating the need to duct bleed air throughout the aircraft. However, X1 used thrust modulation through engine speed, and X2 used fully modulated afterburner from idle to 100% power. Using the rudder pedals, the ducts could be rotated differentially to produce yaw motions. The control surfaces remained active while hovering

Should an engine fail during hover, the autostabilization system would open the remaining engine of that pair to full power and throttle back the other engines to 50% power each. While the aircraft would settle hard to the ground depending on its weight and altitude, it at least would remain stabilized and controllable. Engine loss at 30 feet would be survivable, 50 feet would be questionable, but from 100 feet there would be adequate time to eject. The probability of an engine fail-

ure was calculated to be once every 15,000 hours, a risk that was considered acceptable.

During the design, and prior to and during flight testing, a free-flight hovering rig was used to investigate hovering characteristics, design the autostabilization system, test failure modes and response to wind gusts, and train pilots by giving them the feel of the aircraft while hovering. The orientation of the engines, cockpit, and landing gear was as similar to the actual VJ 101C as possible. The ratio of thrust to moments of inertia were equal to the actual aircraft so that the motions resulting from the pilot's stick and engine commands in the hovering rig would be very similar to those that would be experienced in the actual aircraft. The test rig also had the capability to swivel the wingtip engines +/- 6 degrees from vertical for yaw control.

Next, the actual VJ 101C was mounted in a captive hover rig. The rig used a telescoping pylon to allow vertical motion and three rotations that let the pilot experience the actual aircraft's pitch, roll, and yaw motions, as well as vertical translations in response to stick and throttle movements. Final development of the stability augmentation system was performed using the captive hover rig. Evaluations of pure manual control, attitude control, rate control, and integrated rate control with attitude control were performed. The purely manual system, without any stability augmentation, was found to be unacceptable. The attitude command system, in which aircraft attitude is directly proportional to pilot control stick position, was selected for further development. The rig also proved to be a good pilot training device. The use of thrust modulation for attitude control was demonstrated to be effective and gave added confidence to the pilots and design engineers as they neared the first flight. These tests began in December 1962, and continued up until the aircraft was readied for flight testing.

The first flight, a hover test, was made by vehicle X1 on April 10, 1963. Hovering test flights were relatively short because X1 could only carry enough fuel for 15 minutes of hovering. The first conventional take-off and landing was performed on August 31, 1963. The narrow undercarriage and

Note the narrow landing gear and the auxiliary air inlet that opened at low speed and high power. Also note the open inlet door for the lift engines. (Deutsches Museum)

The VJ-101 in high-speed level flight. (Deutsches Museum)

high-mounted outboard engine pods (i.e., high center of gravity and high roll inertia) did not present any problems with taxi, take-off, or landing. Thus, the initial, arbitrary cross wind limit of 10 knots was soon raised to higher speeds. The first transition was performed on September 20, 1963, nine months ahead of schedule. The first double transition was performed on October 8, 1963, after a total of only 1 hour and 46 minutes flight time.

Testing of X1 in hover, transition, and conventional flight progressed rapidly. A speed of Mach .8 was achieved in level flight with only 20 percent power in April 1964. Confidence in the new aircraft was high enough that it was taken to the German International Air Show at Hanover in late April of 1964 and it performed transitions for the public. On July 29, 1964, it exceeded Mach 1.08. EWR boasted that the VJ-101 was the first VTOL aircraft to exceed Mach 1, but the Hawker Siddeley P.1127 already had achieved it in late 1961 "by an ample margin."

In other conventional flight tests, X1 demonstrated descents at 90 knots airspeed, 18 degree angle of attack and 4,000 feet per minute rate of descent. There was no indication of a stall, largely because of the end-plate effect of the engine pods. Level stalls at about 140 knots were very clean. Overall, excellent high and low speed compatibility and good handling qualities were observed.

Unfortunately, X1 was lost on September 14, 1964. Following a conventional take-off at an altitude of only 10 feet, it rolled, first in one direction, and then in the other, and then dived into the ground. The pilot ejected, suffering a crushed vertebrae, but eventually returned to flying. Damage was extensive and X1 was scrapped. It was determined that the engines were running normally, and the crash was caused by a malfunction in one of the electronic systems, and not related to any VTOL component.

Test vehicle X2 was in final assembly at the time of the crash, and ground tests started in October 1964. It made its first hover flight on June 12, 1965. As with X1, testing progressed rapidly. X2 made its first conventional flight with afterburners on July 12, 1965, first vertical take-off with afterburners on October 10, 1965, and first double transition on October 22, 1965. X2 was damaged in a hard landing that

broke the landing gear, but it was repaired and returned to service.

While hovering flight proved to be very stable, the VJ 101C suffered from major problems with both types of hot gas ingestion. Near field effects resulted from hot exhaust reflecting off the runway and hitting the bottom of the wing and fuselage. This could produce a potentially dangerous, upsetting rolling moment if the aircraft lifted off in a banked attitude or rolled immediately after a vertical take-off. The hot exhaust also caused the underside to get very hot and damaged the tires. Far field effects result from the hot exhaust reflecting off the runway and traveling outward, then rising and getting sucked back toward the aircraft by the airflow into the engines. This resulted in reduced power because the air going into the engines was now hotter and less dense. Also, with the jet engines so close to the ground, there was major erosion damage to the concrete runway, especially with X2 with its afterburning engines.

During vertical landings, a ground cushion was distinctly noticeable at about 10 feet altitude and considerable power reduction was needed to continue the vertical descent. As hover testing progressed, it was determined that recirculation built within 5 – 10 seconds of reaching full power on the ground, but the aircraft would rise out of it after climbing only 3 – 4 feet. With this knowledge, a jump take-off technique was developed that overcame most of the problems. Normal vertical take-off technique was now to jump quickly into the air to avoid near field effects, then immediately tilt the nose down slightly to begin moving forward to eliminate far field effects and other problems with ground heating and erosion. However, it was determined that a forward speed of at least 29 knots was needed in order to outrun the portion of the hot exhaust gas that was reflected forward.

Another take-off procedure used to limit hot gas ingestion and ground erosion was the rolling vertical take-off. With the engine pods tilted to 30 degrees from vertical, the VJ 101C lifted off after only a 10 foot roll and reached a 50 foot altitude after traveling only 130 feet. Rolling landings with afterburner needed only 165 feet of runway.

Transition from hover and from conventional flight proved to be relatively straightforward. Two switches were located on the throttle to control the pods: one to select the direction of pod tilt, the other a three position switch with center detent to select fast or slow pod rotation. To begin the transition to a hover, the landing gear and flaps were lowered at 220 knots, and the lift engine intake and efflux doors were opened to get the lift engines started. The engine pods were turned from horizontal to 45 degrees at 165 knots and power was reduced to 50 percent. Thrust from the forward lift engines was increased as a function of the pod angle by a mechanical linkage as they rotated toward vertical. Level flight was maintained down to 90 knots, at which time a conventional landing could be made. To continue with a vertical landing, thrust was increased and the nose raised to decelerate to 50 knots, then the pods were turned to vertical and the nose raised to bleed off remaining airspeed. The transition was completed as the aircraft slowed to zero forward speed. The entire process took about 90 seconds and one mile. The transition from hover to conventional flight essentially was the reverse.

In mid 1963, the combat roles amongst the NATO partners shifted and the interceptor role was assigned to the WWII allies. The role of attacking enemy aircraft on their airfields was given to the Luftwaffe. This changed the primary aircraft requirement to that of a low altitude strike fighter with supersonic high altitude capability. This new mission made high fuel capacities necessary for low altitude flight over long distances and led to the VJ-101D design. In spite of the similar designation, the VJ 101D would have been a completely new design capable of speeds of 1,500 miles per hour.

The propulsion system consisted of 2 fuselage mounted lift/cruise engines, placed side by side and each equipped with a variable thrust deflector, and five lift engines in tandem. The wingtip mounted engine pods were deleted. The lift/cruise engines would have been Rolls-Royce/MAN RB.153-61 turbojets, each producing 6,850 pounds thrust, or 11,750 pounds with afterburner, and the lift engines would have been Rolls-Royce RB.162s producing 5,500 pounds thrust each.

Note the massive fuselage and tiny wing of the VAK 191B in flight. It was built for a vertical take-off, nuclear strike mission. (Deusches Museum)

Studies indicated that the thrust to weight was barely adequate and that the basic design could not accommodate any weight growth. The German Ministry of Defense canceled the VJ 101D in mid 1964 in favor of the VAK 191B, which was considered to be cheaper and less risky.

As explained earlier, the need for an interceptor with VTOL capability had already ended by the time the VJ-101C entered flight test. Indeed, NATO lost interest in VTOL attack aircraft altogether by the mid 60s, leaving the two VJ 101Cs as research aircraft. X2 remained in service as a flying test bed for advanced V/STOL concepts and flight control system development. From an operational standpoint, the VJ 101C's basic design showed severe limitations in tactical utility due to its small fuel capacity, minimal payload, and serious other operational constraints. Flight testing ended in June 1971 when the lifetime of major components ended. X2 currently is on display at the Deutsches Museum in Munich, Germany.

VFW VAK 191B LIFT + LIFT-CRUISE VTOL (GERMANY)
System: Built-from-scratch prototype
Manufacturer: Vereinigte Flugtechnische Werke (VFW)
Sponsor: German Air Force
Time Period: 1961 - 1972
Mission: Prototype VTOL nuclear strike fighter
The Story:
The German Air Force began developing the VAK 191B at about the same time the British were developing P.1127, which evolved into the Harrier. Work started in September 1961, following issuance of the NATO Military Basic Requirement No 3. This specification called for a single seat, close support and reconnaissance tactical fighter that could incorporate V/STOL technology and have a combat radius of 250 nautical miles, cruise at Mach .92 at 500 feet altitude, and carry a 2,000 pound store. It also specified a loiter time of 5 minutes in the combat area, vertical take-off and landing, a 10 per-

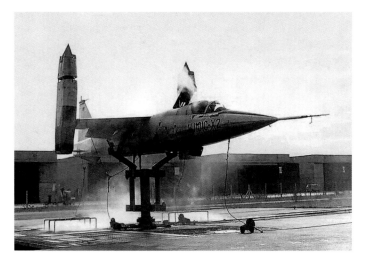
The thust of the VJ101's engines is plainly evident in this photo.

cent fuel reserve, operating in bad weather from minimally prepared operating sites, and using internal power and little other ground support equipment. Estimated cost for the airframe development and prototype production was $56.7 million. The German Air Force anticipated a production run of at least 200 aircraft.

While the public acknowledgment talked of a tactical aircraft to replace the Fiat G.91, the specification, in reality, was for a strike aircraft to carry a single nuclear weapon at high speed and low altitude. The dominance of this single mission eventually proved to be the VAK 191's demise, as the predominate philosophy by the late 1960s shifted to using multi-role aircraft.

The name, VAK 191B, derived from the method in use at the time for designating aircraft. VAK stands for Vertikal Startendes Augklärungs und Kampflugzeug (vertical take-off reconnaissance and fighter aircraft). The 191 designated a successor to the G.91. The B designated that this design was the second of the four aircraft considered in the study. Actually, there were only three designs proposed, the fourth being the British P.1127 that served as a baseline against which to make comparisons. Thus, the three designs proposed in spring 1962, one each from Focke-Wulf, EWR, and Fiat, were:

VAK 191A: P.1127 Mk.2 (reference aircraft, not an actual competitor)
VAK 191B: Focke-Wulf 1262
VAK 191C: EWR 420
VAK 191D: Fiat G.95/4

The EWR and Fiat proposals each proposed using four lift engines and a cruise engine mounted horizontally. The winning design from Focke-Wulf, called the FW 1262, incorporated a center-mounted Rolls Royce/MAN RB.193 lift/cruise engine that could provide about half of the vertical thrust. A pair of RB.162 lift engines, one in front of and one behind the lift/cruise engine, provided the remaining vertical thrust. The design philosophy behind the winning configuration balanced two opposing conditions. First, using only a single lift/cruise engine means that it has to be sized for the vertical take-off. As a result, it would operate at a lower power, and therefore less efficient thrust range, during most of the mission. Second, adding lift engines to augment the lift/cruise engine only during vertical take-off and landing means that they will in-

VAK-191 on display at the Deusches Museum. (Ulrich Seibicke Photo)

crease the size of the fuselage and be dead weight during most of the mission. The optimal solution had to minimize the penalty associated with each of these configurations. The designers figured that the two RB.162 lift engines would add about 556 pounds. On the other hand, a lift/cruise engine that could produce 21,000 pounds of thrust for vertical take-off probably would weigh at least 556 pounds more than the RB.193. This configuration and the approximately 50-50 split of power between the lift/cruise and lift engines proved optimal for the VAK 191B's specified mission, but ultimately it proved too inflexible as the mission evolved. However, it would have been politically undesirable at the time to select a single engine configuration, as in the P.1127.

The overall configuration was a high-wing, tandem landing gear aircraft with small outrigger wheels at the wing tips. The landing gear, along with low pressure tires, permitted touch down sink rates of 12 feet per second. Fuselage length was 48 feet, with a wingspan of 20 feet (this was longer than the P.1127, but a much smaller wing). The airframe and control surfaces were made of aluminum, aluminum-copper alloy, and aluminum-zinc alloy. The horizontal tail was a single piece, all-moving design. To ease remote site servicing, the engine was removed and replaced from under the fuselage (as opposed to the P.1127, in which the engine must be inserted through the top, first requiring removal of the wing—a job which proved to be a maintenance officer's nightmare). Internal fuel capacity was 687 gallons, stored in seven tanks in the center fuselage and one in the rear fuselage. The two lift engines tilted 12.5 degrees from vertical, thus producing a small amount of forward thrust. Diverter doors on the bottom of the fuselage deflected exhaust from the lift engines aft. This produced enough horizontal thrust to maintain level flight at low speed in the event of a lift/cruise engine failure. The rear engine diverter door also could be opened to act as a speed brake. The four nozzles on the lift/cruise engine could be vectored up to 10 degrees forward of vertical to compensate for the forward thrust developed by the lift engines during hovering. The position of the lift/cruise nozzles also were

A VAK-191 test rig is shown in hover mode. (VFW Photo)

coupled to the flight control system, so that they could divert the thrust downward during maneuvering flight in order to help reduce the turn radius. A Martin-Baker Mk 9 seat, capable of ejecting the pilot at zero altitude and zero airspeed, was installed.

The remarkably small wing, with a hefty 12.5 degrees of anhedral, was sized for the original mission. It was just big enough to produce the lift required for high speed, low altitude, and limited maneuverability. To give the pilot as smooth a ride as possible in the turbulence associated with low altitude flight, the wing sweep angle and airfoil shape were selected carefully so the wing would produce only a small change in lift as the angle of attack increased. A trailing edge flap, augmented by ailerons that could be drooped downward together, allowed slow enough speeds for conventional take off and landing.

The structure was primarily of aluminum, although titanium was used in areas requiring greater heat tolerance. Aluminum sandwich and composite materials also were used.

The primary payload area was a spacious equipment bay in the bottom of the fuselage. This allowed the VAK 191B to carry an assortment of items, including bombs, reconnaissance equipment, a gun, rockets, guided missiles, or additional fuel.

During hovering, the two lift engines, each producing 5,600 pounds of thrust, were controlled together by a single throttle and could not operate differentially to provide pitch control. Attitude was maintained during hovering by small thrusters located at the wing tips and fuselage extremities, powered by bleed air from all three engines. In the event of the loss of one of the lift engines, the other would automatically shut down to prevent a dangerous thrust imbalance. Naturally, this called for an extremely reliable monitoring system with zero tolerance for improperly detecting an engine loss. A hard, but at least level, landing would result.

The lift/cruise engine produced 10,000 pounds of thrust, which exhausted through two vectorable nozzles on each side of the fuselage. Air intakes for the lift/cruise engine were of about conventional size for a fighter/attack aircraft of that size and maximum speed. That was allowed, because the lift/cruise engine produced only half of the total vertical thrust. By comparison, the P.1127, in which the single engine produced all thrust, required sizing the inlet for full power at zero forward airspeed. The smaller inlets on the VAK 191B produced significantly less drag. Diverter doors to direct air into the lift engines, as used on the Russian Yak 38, proved unneeded. Even though the lift engines were oriented almost perpendicular to the airstream, the airflow around the fuselage provided adequate flow through the engines for starting in flight.

Working together, the three engines could produce about 21,000 pounds of vertical thrust. At the specified maximum vertical take-off weight of 17,600 pounds, this provided a 20 percent thrust margin for inevitable future weight growth.

The design of the VAK 191B emphasized the development of several new technologies. These included onboard system checkout and fault reporting; high pressure, 4000 pounds per square inch hydraulic system to allow smaller hydraulic components; a tail-mounted auxiliary power unit to provide electricity, hydraulic pressure, and compressed air for ground testing; and engine starting (to relieve the requirement for external power). The VAK 191B was one of the first aircraft to feature redundant fly-by-wire flight controls, in which the control stick motions are sensed electronically and electrical signals are sent to control the hydraulically powered control surfaces. The electronic flight control system included a self-monitoring feature that continuously checked its own operation and notified the pilot of any faults. In the event of a complete failure in the electronic flight controls, there was a backup mechanical flight control system that used mechanical linkages to open and close hydraulic valves to move the control surfaces.

To make hovering easier, the control stick worked with the fly-by-wire flight control system to produce an attitude command system. When the pilot's hand was off the stick, the system commanded a level attitude. If the pilot moved the stick either longitudinally or laterally, the flight control system would produce a pitch or bank attitude proportional to the stick displacement. This mechanization would greatly reduce the pilot's workload, especially during vertical landings in poor weather, and make it much less likely for the pilot to get the aircraft into a dangerous attitude inadvertently. Overall, this made the hover maneuver much safer. In wing-borne flight, the flight control system changes to a rate command system, in which a given stick deflection produces a certain roll or pitch rate. When the pilot releases the stick, the aircraft stops rotating in whatever attitude it was at when the stick was released.

A hovering test rig, which was little more than an open girder fuselage, was built to help develop the control laws for the flight control system. It used five Rolls Royce RB.108 lift jets. Three of the jets were mounted in the fuselage center to simulate the lift/cruise engine, and one fore and one aft to simulate the lift jets. It was equipped with a Martin-Baker Mk4 ejection seat.

VAK-191 on display at the Deusches Museum. (Ulrich Seibicke Photo)

The 300 horsepower Humbold-Dentz T112 auxiliary gas turbine, installed in the very aft end of the fuselage, started with its own electric motor and battery. It could provide electricity for operating all electrical systems, plus cooling air and hydraulic pressure for starting the lift/cruise engine. This made the VAK 141B independent of the need for ground power for servicing and starting. The auxiliary gas turbine also could be started in flight to provide emergency electrical and hydraulic power.

By June 1964, the United States Air Force began to take serious notice of the VAK 191B, and five months later signed an agreement for U.S. participation in developing the avionics. Fiat of Italy also expressed a great interest, and by July 1965 signed on as a partner to develop the aircraft jointly. By this time, it was decided that the initial development program would build three single-seat prototypes, three two-seat prototypes, plus a static-test airframe. Germany would remain the design leader and would pay 60 percent of the development cost. In order to improve its overall resources, Focke-Wulf joined with another company, Weser-Flugzeugbau, to form VFW (Vereinigte Flugtechnische Werke). Under the agreement between VFW and Fiat, VFW would build the center fuselage and wing center section. Fiat would build the fuselage extremities, wings, and tail assemblies. The lift/cruise engine, the RB.193, would be developed specifically for the VAK 191B by Rolls-Royce in England and MTU (Motoren-und-Turbinen Union) in Germany. The RB.162 lift engines would be virtually off-the-shelf items and would be purchased directly from Rolls-Royce. To keep the overall design simple, the lift engines would have no accessories, and would start only by using bleed air from the lift/cruise engine. Because they spooled up to operating thrust very quickly, they were started just seconds before a vertical take-off, helping to prolong their life and avoid ground erosion and ingestion of flying debris.

About 8,900 hours of wind tunnel testing were performed to refine the aircraft's shape and determine the aerodynamic characteristics. This was composed of 4,400 hours in subsonic flight, 2,000 hours of VTOL and transition flight, 500 hours investigating transonic characteristics, and 2,000 hours

investigating the recirculation of exhaust gasses. Manned, fixed base ground simulators that incorporated actual aircraft electronic and hydraulic components were used to develop the flight control system.

About the time the program really started to come together, it started to fall apart. The German Air Force abandoned the need for a pure nuclear attack aircraft in favor of a much more maneuverable fighter, as the NATO policy shifted from massive nuclear retaliation to flexible response. Escalating costs the next year resulted in abandoning the two-seater development and building those airframes as single-seaters. Further differences between plans for operational use led to Italy's withdrawing from the program in August of 1967. However, Fiat remained in the program as a subcontractor to VFW in order to retain the technical expertise Fiat had developed up to this time. In a further blow, the prototype development was cut from six to three aircraft. As a further sign of the inflexibility of the basic design, the American and German governments began looking at a new project, the AVS (Advanced Vertical Strike) fighter. By 1968, the VAK 191B was reclassified as an experimental program to develop and refine new technologies for other aircraft, particularly in the area of flight-control systems and fly-by-wire.

The prototype was rolled out of VFW's facility at Bremen in April 1969. Engine operation and other systems were checked out over the next seventeen months. As part of the ground test program, the aircraft was mounted on a three-axis pivot, the engines were run, and the aircraft's reactions to the pilot's stick motions were measured. Recirculation tests confirmed earlier wind tunnel predictions that ingestion of exhaust gasses during hover would be minimal.

The first prototype finally flew on September 10, 1971, making a vertical take-off and landing. This was followed shortly after by the first flight of the second prototype on October 2, and of the third prototype in early 1972. Flight tests in hovering and at forward speeds up to 80 knots were performed at Bremen. The lack of any significant amount of exhaust reingestion into the lift engines was confirmed by measuring temperatures at the engine inlet (a temperature increase would have indicated hot gasses flowing around the aircraft). Facilities at Bremen soon proved inadequate to perform transition flight tests, so the prototypes were transferred to the Luftwaffe's test facility at Manching, about 350 miles away. For the trip, the VAK 191Bs were transported underneath a Sikorsky S-64 helicopter, rather than being flown or disassembled and transported. This permitted the aircraft to be moved intact, minimizing the costs and risks associated with re-assembly.

Flight testing continued at Manching. The first transition from vertical to horizontal flight was performed on October 26, 1972. One interesting characteristic during vertical landing was noted. Because of the fountain effect of the vertical thrust being deflected off the ground and hitting the underside, the VAK 191B demonstrated an increase in lift as it settled to within 10 feet of the ground. During a steady vertical descent, the aircraft would stabilize a few feet in the air, and the pilot had to retard the throttle in order to land.

Other flight testing provided mixed results. Conventional take-off speed proved to be nearly 230 knots due to the small wing. Rotation was difficult because of the landing gear arrangement and the small tailplane. Short take-off performance was also poor, for the same reasons. The hot exhaust pointed at the ground caused serious ground erosion, prohibiting vertical take-offs from concrete runways. VTOL operations had to be limited to special metallic surfaces. Hot gas ingestion also proved to be a problem when hovering in ground effect. The transition speed from vertical to wing-borne flight of over 200 knots was considered too high, and the lift engines produced excessive drag when accelerating to the transition speed. However, when out of ground effect, the fly-by-wire flight control system provided for good precision and low workload for the pilot. Handling qualities in wing-borne flight were satisfactory, up to 300 knots, the highest speed the VAK 191B ever achieved.

Efforts to reinvigorate the production program never produced any results. A VAK 191B Mk 2 was proposed that would emphasize high maneuverability and good weapons flexibility. This aircraft would have a 50 percent larger wing to improve its turning radius and conventional take off and landing performance. Other changes would be a 30 percent thrust increase to the lift/cruise engine and a 5 percent increase to the lift engines, four wing hard points and six fuselage hard points to allow external stores to be carried, and a new avionics system to provide better battlefield needs. In late 1971, VFW-Fokker (as it was known by then) teamed with Grumman to propose a Mk 3 version, in response to the U.S. Navy's call for a low-risk sea control aircraft (which ultimately was awarded to the North American XFV-12A). The Mk 3 design called for even larger lift engines, a 20 percent increase in internal fuel, an 18 feet per second vertical speed at touchdown, and pitch control using power modulation from the lift engines. Both of these efforts proved futile. The German Ministry of Defense finally terminated the last aspects of the program in November 1972.

The VAK 191B was an unfortunate example of how an aircraft with a very limited mission can become very vulnerable to changing military needs. Its design proved too specialized to its original mission, thus lacking the growth potential to expand into a broader mission.

DORNIER Do31 LIFT + LIFT CRUISE VTOL (GERMANY)
System: Built-from-scratch prototype
Manufacturer: Dornier-Werke GmbH
Sponsor: German Ministry of Defense
Time Period: Mid-1960s - 1970
Mission: VTOL transport
The Story:
The Dornier Do31 was a prototype/demonstrator for what was hoped would be the world's first VTOL transport. The German government initiated the development effort. For a while, it had the interest of the British government, as well.

Dornier performed studies during the early 1960s to determine what type of aircraft could meet future V/STOL transport needs. They eliminated the helicopter because of its poor speed, and the tilt wing because of its mechanical complexity. They further eliminated any propeller drive concept because of reduced efficiency at high altitude. They finally selected an all-jet configuration, but with the provision that the aircraft spend an absolute minimum amount of time hovering in this regime that is most inefficient for an all-jet V/STOL aircraft.

The German Ministry of Defense conducted the NATO NBMR-4 competition. They awarded a contract for building the three Do31 prototypes to Dornier-Werke GmbH. The de Havilland Aircraft Company at Hatfield, England, performed engineering work, and the Vereinigte Flugtechnische Werke GmbH and Hamburger Flugzeugbau GmbH in Germany fabricated assemblies under subcontracts.

The operational mission called for the Do31 to carry 3 to 4 tons of cargo for two 270 nautical mile legs without refueling. The first take-off was to be STOL, clearing a 50-foot obstacle within 500 feet, but the second take-off and both landings were to be VTOL. The mission further required the Do31 to have adequate thrust and attitude control to continue a

Dornier Do.31. (Dornier Photo)

take off or make a controlled landing over a 50-foot obstacle following the loss of any one engine.

The Do31's basic configuration was a high wing transport with a circular cross section fuselage and a cruciform horizontal tail. The wing was swept back 10 degrees and had no dihedral. Propulsion consisted of a 15,500 pound thrust Bristol Siddeley BS53 Pegasus B.Pg5-2 turbofan lift/cruise engine mounted under each wing and four 4,400 pound thrust Rolls-Royce RB.162-4D lift engines in a pod at each tip of the stubby wing. The Pegasus was not the ideal choice, being larger and heavier than really required. However, at that time there was no other choice for an engine that could function for both cruise and lift. The main landing gears retracted into the Pegasus nacelles. The crew entered through an access door on the forward left side of the fuselage, and there was a cargo door that opened from the rear. The maximum vertical take-off weight ranged from 43,000 pounds at 86°F and 5,000-foot altitude, up to 60,000 pounds at 32°F and sea level. Cruise speed was 350 knots, service ceiling was 35,000 feet, and maximum climb rate was 3,740 feet per minute.

The maximum design Mach number was 0.7, relatively low for a jet transport. This resulted from the lift engine pods reducing the critical Mach number. (Critical Mach number is the cruise Mach number at which localized supersonic flow begins somewhere on the airframe, causing large drag increases for very small cruise speed increases. An improvement could have been realized by increasing the wing sweep, but this would have added an unacceptable amount of weight.) However, this low cruise speed was not a significant limitation because of the short flying distances planned.

Pointing the Pegasus nozzles aft allowed conventional take-offs to be performed. Pointing them in an intermediate position allowed short take-offs. The eight lift engines enhanced short take-off performance because they slanted 15 degrees aft of vertical. In addition, each lift engine also had a vectorable nozzle that could rotate 15 degrees fore or aft of the engine centerline.

Wing-mounted control surfaces consisted of a split flap that extended from the fuselage to the lift/cruise engine pod, and an aileron from the lift/cruise engine pod out to the lift engine pod. The ailerons were very effective and could produce roll rates up to 125 degrees per second. At one public

demonstration, the Do31 even performed a barrel roll in front of the crowd!

Cockpit controls consisted of a conventional center stick and rudder pedals, a throttle for each Pegasus engine, and a single nozzle lever to control all Pegasus nozzle angles. A single throttle controlled all eight lift engines for height control. The center stick modulated the lift engines for roll control. During hover, four nozzles at the rear of the fuselage controlled pitch attitude. Two pointed up and two down, and were powered by high-pressure bleed air from the Pegasus engines. Yaw control was by differential vectoring of the lift engine nozzles. Having the lift engines so far outboard greatly assisted roll and yaw control.

With so many engines to manage, it would have been very difficult to identify a malfunction quickly. Two banks of eight engine failure lights helped the pilot locate the problem engine. One bank indicated if any engine was not producing proper speed, and the other indicated if any engine was overheating. The pilot could shut down the errant engine simply by pressing the illuminated light. To maintain a level hover, an opposite engine also had to be shut down.

Stability augmentation assisted in all three axes for hover and transition. The only flight control mode developed was an attitude command system, in which the pilot positioned the aircraft at a desired attitude and the system then maintained that attitude.

Given that the Do31 had ten engines producing a total of 66,200 pounds of thrust in hover, the flight control system could compensate for the loss of any one engine. If a lift engine failed, the system immediately adjusted the remaining throttles to control roll, and the collective throttle to control height. If one of the Pegasus engines failed, a quarter of the total thrust would be lost. The system would run the lift engines on the failed side at an emergency thrust setting, the good Pegasus at maximum thrust, and the remaining lift engines on the good side at less than maximum thrust to control roll.

Note the open doors on the lift-engine pods on the Dornier Do.31 during vertical flight. (Deutsches Museum Photo)

Extensive testing using models preceded the design. Several complete and partial models were tested in wind tunnels at Stuttgart, Göttingen, and Brunswick, as well as in the Dornier research facility at Immenstaad. In addition, NASA's Langley Research Center conducted free-flight tests using a scale model propelled by compressed air. Several flying test rigs were used to aid the design of control techniques and the stability augmentation system for hovering flight. These varied from a simple cruciform-shaped truss structure with a lift engine at each "wing" tip, to a large test rig that actually resembled the Do31. It featured an enclosed cockpit, an actual wing with Pegasus engines, and three RB.162s in each wing tip pod (fewer lift engines were needed because of the lower weight). Most of the fuselage even was enclosed to better replicate recirculation problems while in ground effect.

One of the most successful uses of the test rig was to determine the optimal vertical take-off configuration. In ground tests, the Pegasus nozzles gradually were rotated toward the vertical position in 5 degree increments, looking for the point at which the engines would suck in the exhaust gases. All went well up to 85 degrees. At 90 degrees, lift engine intake temperature rose by 35 to 55°F, indicating serious recirculation. Past 90 degrees, the Pegasus inlet temperatures also started to rise. In addition, loose gravel and concrete chips were blown into the air and promptly sucked into the Pegasus. The best liftoff technique proved to be with the nozzles at 85 degrees and the test rig lifting off with a 5 degree nose up attitude. Once out of ground effect, the pilot could lower the nose and further adjust the nozzles. Adjusting the nozzles to 115 degrees resulted in a level attitude and permitted a three-point touchdown. Test rig flights were flown in 1964, 1965, and 1967, and proved invaluable not only in developing the aircraft, but also in training the pilot.

The German Ministry of Defense and the British Ministry of Aviation joined forces to look for additional users for the Do31. They signed an agreement in April 1964 to investigate expanding the Do31 program to meet transport requirements of both governments. While not a direct partner in the Do31 development, British interest in a follow-on program helped keep the Do31 development alive. No foreign partners ever participated, and all funding for the Do31 came from the German government with some in-house investment by Dornier. By January 1966, the British government pulled out of the study, stating that they had no operational requirement for a military VTOL transport.

Dornier produced three prototypes, designated E1, E2, and E3. E1 lacked the RB.162 lift engines and was used to investigate conventional take off and landing characteristics. It made its first flight on February 10, 1967. E2 was built solely for ground testing and never flew. E3 was built with all engines installed and made its first hover flight on November 22, 1967. It performed its first transition from conventional flight to hover on December 16, and the first transition from hover to conventional flight on December 22. Do31 E3 made the first public flight demonstration at the Hanover Air Show

in May 1968. Only one pilot ever flew the Do31, an American test pilot named Drury Wood who worked for Dornier. Because of limited schedule time and tight development funds, Dornier felt that it was not practical to have more than one pilot involved.

Hover tests indicated very little, if any, problem with reingestion of exhaust gasses. Credit for this went to the 15 degree offset of the lift engines that blew the exhaust to the rear and well clear of the engine inlets. Transitions from hover to conventional flight were performed between 30 and 1000 feet of altitude. However, the transition normally started as soon as the aircraft left the ground, with full wing-borne flight being achieved at 500 foot altitude and 140 knots airspeed in about 15 seconds. Vertical landings started at 140 knots and 600 ft altitude with the Pegasus nozzles rotated to a forward position to give a decelerating thrust while the lift engines were started. Full hover was obtained by reaching 50-100 feet of altitude. Total time from start of transition was 75-90 seconds. The Dornier test pilot reported that the Do31 was easier to hover than any helicopter he had ever flown.

By mid-1969, the Do31 made over 200 vertical take-offs and transitions to conventional flight. Landing transitions were beginning to be studied during early 1970.

The proposed production version, which never made it to the prototype stage, maintained the same overall configuration, but differed in many ways. The useful load would have been 10,700 pounds, and the range would have been 500 nautical miles. Changes included:

• Moving the horizontal stabilizer to the top of the vertical tail
• Tapering the wing and giving it some negative dihedral
• Moving the main landing gears to pods on the sides of the fuselage
• Replacing the Pegasus lift/cruise engines with unspecified 8,800 pound thrust turbofans fitted with exhaust deflectors
• Moving the lift engine pods inboard to about mid-span
• Replacing the four lift engines in each pod with five lift engines of about 5,500 pounds thrust each
• Adding pitch thrusters at the nose
• Adding the capability to replace the lift engine pods with auxiliary fuel tanks for non-VTOL operations.

The German government finally terminated the Do31 program in April 1970. The primary reason given was that the

The wingtip lift engines of the Do.31 are clearly visible from this angle. (Dornier Photo)

The wingtip-mounted engine pod on the Do.31 is clearly visible in this photo. (Dornier Photo)

large drag and weight of the engine pods reduced the useful payload and range to unacceptable levels. But most likely, the Do31 was an aircraft than never really had a mission and probably never had the government support it needed. Because of extremely limited funds, ongoing threats of program cancellation, and close government scrutiny, testing progressed very cautiously. Contrary to the way most research or development programs are conducted, Dornier made very few improvements. Once they demonstrated that the aircraft could fly, no changes that potentially could improve performance or handling were made, probably out of fear that a single accident would certainly kill the program. The only changes made were to improve safety deficiencies.

It is also worth taking a look at the primary transportation systems already in place. Europe had an excellent railway system that provided fast, inexpensive, and punctual service. And, the basic infrastructure for the railway system long had been paid for. Also, most major cities had airports that were not unreasonably far from the city center. Thus, there really was no European market for a commercial VTOL transport. Today, the Do31 E3 prototype is on display at the Deutsches Museum in Munich, Germany.

YAKOVLEV YAK-38 LIFT + LIFT CRUISE VTOL (SOVIET UNION/RUSSIA)
System: Built-from-scratch production aircraft
Manufacturer: Yakovlev Design Bureau
Sponsor: Soviet/Russian Navy
Time Period: Mid 1960s - 1993
Mission: VTOL fighter for fleet defense, reconnaissance, and anti-ship strike
The Story:
The Soviet Yakovlev Yak-38, NATO code named Forger, was one of only a very few V/STOL aircraft ever to go into production and become operational. Its primary roles were fleet defense, reconnaissance, and anti-ship strike. Western observers first saw Yak-38s on July 18, 1976, on the anti-submarine cruiser *Kiev* when it was traveling through the Mediterranean en route to the Soviet Northern Fleet base at Murmansk in the Arctic Ocean. The *Kiev* carried a developmental squadron of ten aircraft, designated the Yak-36M at the time.

In a clean configuration, the Yak-38 could obtain a top speed of Mach 1.35, but the maximum level speed at sea level was 620 knots, or Mach 0.94. The intended payload was 2,200 pounds of ordnance, including air to surface missiles, bombs, and ammunition. Three different production versions were built: the original single seat version, a two-place trainer version, and an upgraded single seat version. The operating weight for the original version was 16,500 pounds. A total of 231 Yak-36s were built, including four prototypes, by the time production ended in 1988.

The Yak-38's development and introduction into service presupposed two premises that eventually proved to be false. First, the smaller, 40,000 ton aircraft carriers built for V/STOL aircraft would be cheaper to operate than full sized carriers. Second, V/STOL aircraft quickly would catch up with conventional aircraft in terms of payload, performance, operating envelope, and economics.

Note the folded wing on this Yak-38 on display at a museum in Russia. (Paul Nann Photo)

The basic configuration of the Yak-38 was two lift engines located in the fuselage just aft of the cockpit, plus a single lift/cruise engine. The overall length was 50 feet 10 inches, 58 feet for the trainer version, and a 24 foot wing span. The lift/cruise engine had two vectoring nozzles, one on each side of the fuselage, that rotated from horizontal to just forward of vertical. Fuel was carried in fuselage tanks located forward and aft of the lift/cruise engine. The canopy was hinged on the right side.

The shoulder-mounted wing was swept back 45 degrees and had a small amount of anhedral. For shipboard storage, the wings folded to reduce the span to 16 feet. A single-slotted Fowler flap was mounted at the trailing edge of the fixed, inboard section, and an aileron was mounted on the outboard section. There were no leading-edge flaps or slats. All ordnance mounted on two pylons under the fixed section of each wing, and there was no internal armament.

Attitude control during hover was by reaction jets, with upper and lower slots on each wing tip for roll and on both sides of the tail cone for yaw. Pitch control was by varying the thrust of the lift engines. A Zvezda K-36VM zero-zero automatic ejection system activated if certain limits of attitude, angular acceleration, and/or rate of descent were exceeded. Loss of thrust in any of the three engines could cause an immediate and violent pitching moment, necessitating the automatic system. A steerable nozzle on the seat's rocket motor provided thrust vectoring when ejections were made at low altitude or at unusual attitudes. In 15 years of operational service, over 15 percent of the Yak-38s were lost in accidents, with 13 manual ejections and as many as 20 automatic ejections.

Two Rybinsk RD-36-35FVR lift engines, producing 6,722 pounds of thrust each, were located immediately behind the cockpit and were inclined so that their thrust was pointed 13 degrees rearward.

One Soyuz Tumanskiy/Khatchaturov R-27V-300 turbojet of 13,444 pounds thrust was mounted in the center of the fuselage. A hydraulically actuated vectoring nozzle was located on each side of the fuselage just aft of the wing trailing edge. The lift/cruise engine was modified from the basic R27F2M-300 used on the MiG-23, with the afterburner and nozzle replaced by the rotating nozzles, each capable of rotating through a 100 degree arc in 8 seconds. None of the engines had afterburning. The lift engines were started by bleed air from the lift/cruise engine. A single door hinged on the aft side covered the lift engine inlet bay. To prevent too much air from being deflected into the lift engines, 16 spring-loaded louvers on the inlet door opened to allow excess air to slip past. Two side-hinged doors were mounted on the bottom of the fuselage to cover the exhaust bay.

The first prototype flew in 1971. Given this date, development must have started in the mid to late 1960s. During development, a full scale fuselage mock-up with the propulsion system was carried aloft in the bomb bay of a Tupolev Tu-16 Badger bomber to test engine starts. The fuselage was lowered into the air stream during flight. Four prototypes were built as test vehicles, designated VM-1 through VM-4. VM-1 was used for propulsion system ground testing at TsAGI, the research facility just outside of Moscow. VM-2 and VM-3 were the primary flight test aircraft, and VM-4 was used for final acceptance testing.

The first flight of a Yak-38 was performed on January 15, 1971, using prototype VM-2 in a conventional take off and landing. The first hover followed on September 26, 1971, and the first double transition was on March 20, 1972.

A significant suck down problem was observed in hover. This resulted from high speed gasses reflecting off the runway, blowing against the bottom of the aircraft, and producing downward lift as it flowed around the side of the fuselage. This was overcome by adding ventral strakes on the bottom of the fuselage to capture what is called the fountain effect and negate suck down. Fountain effect is when the exhaust reflected off the runway hits the bottom of the aircraft and simply pushes it upward. Two other significant problems found during flight testing were that the lift/cruise engine's installed thrust was less than expected, due to being turned through the rotating nozzles, and drag was higher than expected. These two shortcomings limited the Yak-38 to subsonic speeds with any useful payload. At some point fences were added to the top of the fuselage on either side of the lift jet inlet. Presumably, these were added to limit ingestion of recirculating exhaust gas during hover.

State acceptance testing began in April 1972 at the Soviet Air Force Research Institute. The Yak-38 made its first vertical landing on a ship on November 18, 1972, touching down on a specially installed platform on the helicopter cruiser *Moskva*. Sea trials were completed in January 1973. State acceptance testing was completed on August 11, 1976.

The "M" designation was dropped upon entering operational service, and the aircraft was designated simply the Yak-38, although one source referred to it as the Yak-38MP, for Morskoy Palubnyi, or maritime carrier-borne. Interestingly, it was cleared only for vertical take-offs and landings. Conventional and short take-offs and landings were not permitted

The Yak-38 carried two lift engines in the fuselage just aft of the cockpit, plus a single lift-cruise engine. (Paul Nann Photo)

due to lingering control software and engine reliability concerns.

The Yak-38 proved to be very stable during vertical take-off and landing. Following a vertical take-off to about 15-20 feet, the pilot started the transition by lowering the nose to 5 degrees below the horizon and maintained altitude as the aircraft accelerated. Upon reaching 30–40 knots forward speed the nose was rotated to 5 degrees above the horizon and the vectoring nozzles were rotated aft.

Landing began by descending and slowing until approaching the stern of the carrier at 100 feet above the water. The stern was crossed at about a 5 knots closure rate and at about 35–45 feet above the flight deck. The final motion was to flare gently to a hover and descend vertically.

Operating limits plagued the Yak-38 throughout its operational life. When the *Kiev* deployed to the South Atlantic, additional weight restrictions were imposed because the elevated temperatures and high humidity caused the lift engines to lose thrust. This resulted in the maximum take-off fuel being restricted to less than 3,000 pounds, nearly half of that being used just for the vertical take-off and landing. The net result was that the low altitude operating radius was only 40 nautical miles. Diverting a portion of the pilot's oxygen supply into the engines finally restored their lifting power.

The Yak-38 deployed to Afghanistan for operational trials during spring 1980 and exhibited similar problems. High ambient temperatures and mountainous terrain limited take-offs to very early in the morning, and even then, with reduced payload and range.

System improvements eventually allowed some of these operating limitations to be reduced. VM-4 was used during December 1978 and January 1979 for additional development tests for conventional take-offs and slow speed landings of about 30 knots. An algorithm was developed which allowed the flight control system to rotate the nozzles in an optimum manner during take-off and landing rolls. The Yak-38 finally was cleared for STOL operations in late 1980. Dozens of rolling take-offs and landings were performed on the *Minsk*. It was found that a 360 foot rolling STOL take-off, even on a 90° tropical day, used 600 pounds of fuel, compared to 800 pounds required for a vertical take-off. Likewise, short rolling landings cut landing fuel in half, to 275 pounds from

Note the open door on the top of the fuselage of this carrier-based Yak-38.

530. These fuel savings resulted in more than tripling the maximum sea level operating radius from 40 to 140 nautical miles, and allowing an increased take-off weight to 24,910 pounds.

Development of the two seat trainer version, the Yak-38U, NATO designation Forger-B, began in 1971. A second cockpit was added forward of the normal cockpit, and a four foot fuselage plug was added to the rear fuselage to maintain the center of gravity. The Yak-38U was 7 feet longer than the single seat model, and the empty weight increased by about 2,000 pounds.

The maximum vertical take off weight was limited to less than 22,050 pounds, achieved by reducing the fuel, and the maximum for short take-off was 25,795 pounds. Although the Yak-38U was intended to have a limited combat capability, the radar and hardpoints eventually were deleted to save weight.

The final Yak-38 version, again designated the Yak-38M, was started in 1981. The RD-36 lift engines were replaced with the more powerful RD-38, producing 7,171 pounds of thrust each, and the lift/cruise engine was replaced with the new Soyuz R28V-300 producing 14,770 pounds of vertical

Dorsal strakes on either side of the left engines help to limit injestion of hot exhaust during hover. (Paul Nann Photo)

thrust. The maximum vertical take-off weight was increased to 24,910 pounds, the same as the previous version for STOL operations. A steerable nose wheel and provisions for 160 gallon under wing fuel tanks were added. Factory testing was performed between December 8, 1982, and June 3, 1983, when production was approved. State acceptance testing continued through June 1985, when the Yak-38M formally entered service.

Yak-38s eventually were operated from four carriers, the *Kiev*, *Minsk*, *Novorossiysk*, and *Baku*. Typical compliment was twelve single-seat aircraft and two two-seaters. Despite the improvements achieved in the Yak-38M, the entire series was removed from service in 1992-1993. The aircraft never completely satisfied the Soviet, and then the Russian Navy's requirement for a supersonic combat aircraft. Its range and payload were significantly less than contemporary conventional aircraft. Also, the collapse of the Soviet Union and the lack of funds to continue development were significant factors. The roles for which the Yak-38 was built were turned over to navalized versions of the MiG-29 and Su-27, operating from full-sized aircraft carriers.

A total of 231 Yak-38 aircraft were built. These included the 4 prototypes, 139 Yak-38s, 50 Yak-38Ms, and 38 Yak-38U trainers. When pulled from service, some of the later Yak-38Ms had flown as little as 12 hours. Some of the older models accumulated as few as 390 hours in 1,100 sorties over ten years of service.

LOCKHEED XV-4B LIFT + LIFT CRUISE VTOL (U.S.A.)
System: Rebuilt XV-4A
Manufacturer: Lockheed-Georgia Company
Sponsor: U.S. Air Force
Time Period: 1966 - 1969
Mission: Lift + Lift Cruise research vehicle
The Story:
When the XV-4A program ended in 1964, the airplane's life proved to be far from over. Lockheed continued looking at other vertical lift systems that could be installed, including fans, direct lift engines, and improved ejector designs. They eventually determined that the best performance could be achieved with lift fans. Direct lift engines came in second, but still far exceeded the performance demonstrated with the jet ejector system.

At about the same time, the Air Force's Flight Dynamics Laboratory (now part of the Air Force Research Laboratory) established the VTOL Flight Control Technology Program to investigate handling quality requirements for VTOL aircraft in hover and transition. This effort and Lockheed's ongoing work proved to be a good match. The Army agreed to give the surviving XV-4A to the Air Force, and the Flight Dynamics Laboratory contracted the Lockheed-Georgia Company in September 1966 to rebuild the aircraft as the XV-4B. The overall program had four phases: to modify the XV-4A into the XV-4B, install a variable stability system to allow the aircraft to demonstrate a wide range of flight characteristics,

XV-4B mounted on a test rig. (Lockheed Photo)

develop and demonstrate the control and display equipment necessary to allow VTOL aircraft to operate in instrument conditions, and finally, transfer the aircraft to the Cornell Aeronautical Laboratory in Buffalo, NY (now the Calspan Corporation), for a long-term effort to develop criteria for VTOL handling qualities (i.e., determine what characteristics make for a well-flying aircraft).

While the aircraft itself looked virtually the same from the outside and retained most of the same structure, almost everything else on the inside was gutted. The largest change was a completely new propulsion system. To insure no shortage of vertical thrust, six modified General Electric YJ85-19 turbojet engines, four mounted vertically and two mounted horizontally, replaced the JT-12 engines and jet-ejector system. Diverter valves on the horizontal engines allowed virtually all of their thrust to be deflected downward to assist the four other engines during vertical flight. Two other engine configurations that retained the JT-12s were also studied, but proved less capable. One was using four vertically mounted J85s and retaining the original JT-12s for horizontal thrust only. The other used two vertical J85s for vertical thrust and two JT-12s with diverter valves so they could provide both horizontal and vertical thrust. While using the six J85s proved to be the most extensive of the modifications, it also gave the best performance and thus was selected.

Given that the XV-4B was a research aircraft that did not require a broad, conventional flight envelope, it was possible to increase the gross weight up to 12,000 pounds while still retaining a large vertical take-off and transition envelope without major structural changes. Maneuver loads were restricted to 3.0g positive and 1.5g negative. Sink rates were limited to 10 feet per second for conventional landings and 13 feet per second for vertical landings. The maximum speed was limited to 260 knots, or Mach 0.53, down from 410 knots/Mach .68 for which the airframe originally was designed.

The XV-4B Hummingbird II VTOL is shown during its rollout at Lockheed-Georgia in June 1968. (Lockheed Photo)

Douglas Aircraft developed the Escapac 1D-3 ejection seat, especially for the XV-4B, based on a system developed for the X-22. The seat permitted safe ejection in all combinations of speed and altitude that were within the aircraft's approved operating envelope. To speed canopy opening for a true zero-zero ejection capability, a "ballistic spreader device" was developed to insure full opening of the parachute in minimal time. During development, four single seat static firings and one dual seat sequenced static firing were performed. As a final test, a dynamic firing of both seats at 80 knots was performed.

Each lift engine had a vectoring nozzle that could rotate from 10 degrees forward to 10 degrees aft to provide fore/aft control during hover and acceleration/deceleration between hover and transition. Care was given to the design of the lift engine exhaust and the landing gear positions so that hot exhaust would not blow directly on to the main tires. As an extra precaution, B. F. Goodrich Co. developed special high temperature tires, including an instrumented tire to provide accurate temperature data.

The flight control system was changed to an electro-mechanical system with the surfaces hydraulically powered. In normal operation, the aircraft was completely fly-by-wire with the mechanical connections disengaged. A triplex fail operational system allowed undegraded performance if a single failure occurred in any one branch of one or all three axes. If a major fly-by-wire system failure occurred, the clutches would engage the mechanical backup system automatically. In the event that both 3,000 pound per square inch hydraulic systems failed, the pilot could still move the control surfaces through the mechanical system, but with much greater effort required. In the event of any control system failure, a warning signal notified the pilot while the system continued to operate. The control system had a set of unused input channels.

These were reserved for the eventual addition of a variable stability system that would allow the XV-4B to produce a broad range of flight characteristics.

The conventional control surfaces were assisted during hover and transition by reaction controls at the nose, tail, and wing tips. These were similar to those on the XV-4A, but used only engine bleed air and operated on a demand basis only.

Each pilot has his own set of cockpit controls, consisting of a stick, rudder pedal, throttle, and collective lever. Spring cams artificially produced the feel forces in the pilots' control sticks and pedals. Although the forces did not vary with airspeed, the spring constant was changed between conventional and vertical flight, making the forces lighter during hover and heavier during conventional flight.

XV-4B under construction at Lockheed-Georgia. (Lockheed Photo)

A view of the side-by-side seating and twin engines of the XV-4B VTOL. (Lockheed Photo)

Stability augmentation in all three axes provided additional damping of oscillatory motions. This made the XV-4B feel somewhat like a small fighter when flown in the conventional flight mode. In V/STOL flight, the system behaved essentially like an attitude hold mode, meaning that after the pilot positioned the aircraft in a given attitude, it tended to remain there.

Each set of throttles consisted of two levers that moved fore and aft for horizontal thrust, and a collective control for vertical thrust. Each pilot also had trim controls for each lift engine and a switch with which to control the diverter valves for the lift/cruise engines. The horizontal engines were started with compressed air from an external source, but the vertical engines were started using bleed air from the horizontal engines.

Two 370 gallon internal fuel tanks were located in the fuselage, one in front of the lift engines, and one behind. An automatic fuel transfer system kept the fuel loads within 100 pounds of each other. For crew comfort, the air conditioning system could maintain 75° F in the cockpit at outside temperatures between 0° F and 100° F using compressor bleed air from the jet engines.

To aid development, numerous ground tests were performed. Engine tests were performed in a specially designed static rig to allow engineers to optimize the inlet designs. Wind tunnel tests were performed at NASA's Langley Flight Research Center and at the University of Maryland. Ground simulations to refine the flight control system were performed at G.E. Controls at Binghamton, NY, and at the Air Force Flight Dynamics Laboratory at Wright-Patterson AFB, OH. Ground vibration tests on the completed XV-4B, in which small shakers were installed on the wing tips, indicated that there would be no potential flutter problems.

This XV-4 flies in formation with a T-33 chase plane. (USAF Photo)

The XV-4B in conventional flight. (USAF Photo)

As a further aid, the XV-4B was suspended under a large truss assembly. The aircraft could operate either locked in place to measure forces and moments generated with the engines running, or with the aircraft free to rotate about all axes and translate vertically so the pilot could get a feel for the response while hovering.

The XV-4B was rolled out on June 4, 1968. In preparation for first flight, engine runs were performed. Some structural heating caused by leakage of hot exhaust from the cruise engines required minor design changes in that area. No problems were noted in taxi and braking tests, nor on nose wheel lift off and drogue chute evaluations. The first flight, made in conventional mode, was performed on September 28, 1968, at Dobbins AFB, GA, adjacent to the Lockheed-Georgia Co. It lasted 30 minutes and reached an altitude of 7,500 feet and a speed of 240 knots. This event was especially significant because these were the first flights in the western world of an aircraft using the new fly-by-wire technology as the primary means of control for conventional flight. A second conventional flight followed two days later.

Conventional flights continued, in preparation for vertical flight tests and then delivery to the Air Force Flight Test Center at Edwards AFB, planned for June 1969. Unfortunately, the XV-4B crashed on March 14, 1969, after entering an uncontrolled roll at 8,000 feet shortly after departing Dobbins AFB. The Lockheed test pilot ejected and suffered only minor injuries. The crash of the second XV-4 ended the program, with the XV-4B never having flown in hover.

YAKOVLEV YAK-141 LIFT + LIFT CRUISE VTOL (SOVIET UNION/RUSSIA)

System: Built-from-scratch prototype
Manufacturer: Yakovlev Design Bureau
Sponsor: Soviet Air Force, then as private Yakovlev venture
Time Period: 1975 - 1992
Mission: Supersonic VTOL strike fighter for fleet defense

The Story:
The Yak-141 was another attempt by the Soviet Union, and subsequently Russia, to develop an operational, supersonic V/STOL fighter aircraft. The Yakovlev design bureau began work on a totally new design when the performance and operational limitations of the Yak-38 became apparent. Yak-141 development started in 1975. The intent was to provide a totally new aircraft with the combat capability of the MiG-29 while retaining the field performance of the Yak-38. The original Soviet designation was the Yak-41 and was used through most of its development. The NATO code name was Freestyle. The designation did not change to Yak-141 until very late in the program's history.

The design of the Yak-141 represented a change in the Soviet design philosophy. The Soviets tended to perform much of their design through prototyping and testing, which is why they always had new designs flying, many of which ultimately proved unsatisfactory and never went into production. While the first flight goal was 1982, construction of a mock-up did not start until the mid-1980s, and first flight did not occur until 1987. This lengthy design time, more typical of western aircraft, was essential in order to avoid another V/STOL aircraft with severely limited capabilities. The time between start of development (1975) and first flight was considerable for a Soviet program, an indication of the difficulty that faced the aircraft designers.

Initially, the primary mission was to be fleet air defense. The Yak-141 was to carry simplified versions of the MiG-29's radar and an air-to-air missile weapons system. By the early 1980s, the mission changed to include a ground attack role, and the designation changed to Yak-41M. This mission change required additional modifications to the radar, addition of an internal GS-301 30mm cannon with 120 rounds of ammunition located under the left air intake, and the ability to carry a wide range of air-to-air and air-to-surface missiles. The Phazotron NO-193 multi-mode radar could detect air, ground, and surface targets, including small fighters up to 50 statute miles away and missile launches up to 68 statute miles away. An assortment of missiles was to be carried, including the medium range air-to-air RVV-AE and R-27 missiles, short range R-73 guided missiles, H-31A and H-35 anti-ship missiles, the H-25 air-to-ground missile, H-31R radar homing missile, anti-submarine weapons, and conventional bombs

The Yak-141 utilized two lift engines mounted just behind the cockpit, and a lift/cruise engine. To provide vertical thrust, the lift/cruise engine's single nozzle rotated downward through 90 degrees, rather than diverting the exhaust through side nozzles. Perhaps the Yak-141's most obvious feature was its twin tail booms, each mounting a vertical and horizontal fin, sticking well beyond the end of the nozzle. This feature resulted from the need to locate the lift/cruise engine's vertical thrust as close to center of gravity as possible, an obvious result of the rotating nozzle design.

The airframe was made from aluminum-lithium alloys and made extensive use of carbon fiber for all the control surfaces, including the all-moving stabilizers, slats, flaps, and wing strakes. Although the structure was designed to 7g, it never was tested to this level before flight testing stopped. Overall dimensions were a wing span of 33 feet, 2 inches, length 60 feet, 0 inches, height 16 feet, 5 inches, and wing area of 341.2 square feet. Empty weight was 25,685 pounds, maximum vertical take-off weight was 34,830 pounds, and maximum short take-off weight was 42,990 pounds.

The cockpit included a Zvezda K-36LV ejection seat with 0-0 capability. A helmet-mounted sight was planned, but never added due to funding cuts.

A triplex digital fly-by-wire flight control system operated in all three axes and in all flight modes. A single digital computer hosted the flight control system, fire control, propulsion control, navigation, and instrument displays.

The propulsion system consisted of two Rybinsk RD-41 lift engines and one Soyuz R-79V-300 lift/cruise engine. While the Yak-38 used modestly modified existing engines for the V/STOL requirement, the Yak-141 engine development was much more extensive.

Development of the Rybinsk RD-41 lift engine began in 1982, and it weighed 639 pounds and produced 9,040 pounds of thrust. Engine development was complete and ready for production when the program was canceled. About 30 engines were built. Installed in the fuselage, the lift engines inclined about 11 dsegrees off vertical. In addition, a vectoring nozzle on each lift engine was capable of deflecting the exhaust forward or aft about 13 degrees. This allowed the thrust to be deflected from 2 degrees forward to assist braking, and up to 24 degrees aft to assist STOL take-off and provide emergency thrust for forward flight. The maximum run time for these engines was only 2.5 minutes, but typical run times for any operation were only about 40 seconds.

The Soyuz R-79V-300 lift/cruise engine used afterburner during conventional and hovering flight. The burner rings could be lit in sequence to modulate the thrust augmentation. Maxi-

mum thrust was 24,250 pounds, which increased up to 34,140 with afterburning. Rather than directing the thrust out the side and then deflecting it either in the vertical or horizontal direction, as does the Pegasus engine used on the Harrier, the R-79 exhausted straight out the back, which minimized thrust loss during conventional flight. For vertical flight, the nozzle rotated to the vertical position. The tailpipe thrust deflector consisted of two rotating pipe sections, similar in concept to a variable elbow used in household heating ducts. It could vector the exhaust over a maximum range of 95 degrees. Deflection from full aft to full forward took 5-6 seconds. Turning losses caused about a 20 percent thrust decrease when operated in the vertical position. The vectoring nozzle was used for hover and short take-off only…it was not used in forward flight for thrust vectoring to enhance maneuverability. It was rated for at least 1,500 cycles. An integral engine and nozzle control system optimized the engine characteristics to the nozzle deflection.

Hover control was by means of puffer jets powered by bleed air from the last two stages of the lift/cruise engine's high pressure compressor. Roll control was by wingtip ejectors. While the first prototype used only thrusters at the tails for pitch and yaw control at low speed and in hover, the sec-

The unique "tuning fork" rear configuration is evident looking straight up at the hovering Yak-141.

ond prototype added nose thrusters and differential engine thrust for better control.

Four airframes were built: two for ground test and two for flight test. Airframe 48-0 was for static and structural tests. 48-1 was for engine integration and ground testing. Of the two flight test vehicles, 48-2, was used to develop and validate the integrated fly-by-wire and propulsion control systems primarily in conventional flight, and 48-3 was used to test low speed and hover flight, and weapon system testing. A two-seat trainer version was being constructed when funding ended the program.

Ground testing was performed at TsAGI, the Central Aerohydrodynamics Institute, located in Zhukovsky, just outside Moscow. Testing included being suspended by a truss assembly over a concrete-lined pit to minimize recirculation effects. Further ground testing also was performed at Yakovlev Bureau facilities. Engine integration and systems tests were performed in 1986 at the Gromov Flight Research Institute, also at Zhukovsky.

Extensive taxi tests were performed using 48-2, followed by a first flight on March 9, 1987, in conventional mode. The lift/cruise engine's nozzle was locked in the horizontal position because this vehicle lacked the flight control software needed to control it. After the flight, cracks were discovered in the lift/cruise engine, resulting in a suspension of flight testing. Limited funding kept the program grounded for two years, until April 12, 1989, when 48-3, the second flight prototype, made a conventional flight. Flight testing in the conventional mode then continued for several more months, reaching a speed of Mach 1.4.

48-3, which was fully equipped for vertical flight operation, performed the first vertical take-off on December 29, 1989, and the first double transition on June 13, 1990. Subsequently, 48-2, which hadn't flown since its first flight, was modified during the winter of 1990-91 with V/STOL flight software and performed its first vertical take-off and landing in April 1991. By that point, the program had accumulated 30 hours of flight testing.

Hot gas ingestion during hover proved to be a problem due to the afterburning main engine. Adding deflector doors to help direct the recirculating hot exhaust away from the engine inlets corrected the problem. Subsequent temperature rise was only 65-85°F.

Rolling take-off tests were conducted during spring 1991. Rotating the lift/cruise nozzle down 6 degrees and pointing the vectoring nozzles on the lift engines to full aft produced the best short take-off performance. The requirement was for a 100-325 foot take-off roll and no more than an 800 foot landing roll. The best short take-off technique developed was the "pointed start" take-off. Using this take-off, the pilot held the aircraft in place using brakes and chocks while he rotated the lift/cruise nozzle to 65 degrees and ignited the afterburner. Upon releasing the brakes, the aircraft would jump over the chocks. Using this technique, the Yak-141 could take off in

as little as 15 feet, and reach 290 knots in about 18 seconds. In addition, over 100 rolling take-offs using a ski-jump also were made.

The preferred take-off technique was the short take-off run of 200-400 feet. This allowed increasing the payload by 2 tons, stretching the combat radius by 1.5-2 times, and doubling the patrol time. A standard concrete runway used for these tests survived over 60 take-offs and landings. The temperature of the runway surface never exceeded 80°C on these runs, and surface erosion was not a problem. The landing run was 660 feet, and the vectoring nozzle on the lift engines contributed to shortening both the landing and take-off rolls.

The Yak-141 set 12 FAI-recognized international V/STOL records in April 1991 for altitudes and time to altitudes with various loads. In registering these records, Yakovlev used the designation Yak-141 on the application, since the Yak-41M designation was classified at the time.

Infrared testing showed the Yak-141 to have a low infrared signature because the tailbooms shielded much of the exhaust nozzle's heat. The maximum demonstrated speed at sea level was 672 knots, or Mach 1.02, and at altitude was 970 knots, or Mach 1.7. Initial rate of climb was over 49,000 feet per minute. Combat radius at best altitude following a STOL take-off was 372 nautical miles with 4,410 pounds of ordnance, and 1,133 nautical miles with 1,000 pounds of ordnance. Flying speed was achieved in 18 seconds following a vertical take-off.

The Russian government canceled the program in August 1991, despite an initial production order from the Russian Navy and the good test results emerging from the flight test program. The reason was that Su-27s' operating from conventional carriers could meet the naval requirements while providing much higher performance. Certainly the shortage of funds in the newly-formed Russian Republic also was a factor.

With the breakup of the Soviet Union and the military budget in shambles, the new Russian government ordered the old Soviet aircraft design bureaus to become independent companies. With this new airplane so close to production, Yakovlev decided to attempt to complete the development as a private venture, now openly using the Yak-141 designation. They attended the 1991 Paris Air Show with a display featuring a video of the aircraft, in hopes of attracting foreign sales or development partners. By this point, the two prototypes had flown over 250 sorties and logged 150 hours.

Continuing the development, Yakovlev funded additional flight tests beginning in September 1991 on the carrier *Admiral Gorshkov* while operating in the Barents Sea. Unfortunately, the test aircraft, 48-3, was severely damaged in a hard landing and fire while landing in bad weather on October 5. The problem was that a small manufacturing fault gave insufficient tension in the throttle lever's detent stop, causing the aircraft to lose power and fall 43 feet to the carrier deck. The pilot ejected manually. The aircraft was repaired, but not to flying status.

In further hopes of generating foreign interest, Yakovlev shipped the remaining flying vehicle, 48-2, to the Farnborough Air Show in September of 1992 for its first public performance. It had accumulated about 75 hours up to this point, and was repainted with the number 141 on the side and promoted as the Yak-141. The estimated cost to complete the development was $385 million, which would include construction and testing of an additional static test airframe and three additional flight test aircraft. Production deliveries could start in 1996. Again, there was some interest, but no deal. The Yak-141 was shipped home, having made its last flight at the Farnborough Air Show. Yakovlev could no longer afford to continue the development on their own, and the program ended in 1992 with a total of about 210 hours being flown on the two prototypes.

Citing NATO's ability to deny use of Iraqi runways during the Gulf War, Yakovlev proposed a land-based version of the Yak-141. It featured a more powerful lift/cruise engine, larger wing and improved aerodynamics, stronger undercarriage, improved brakes, greater internal fuel capacity, more weapons hard points, and a reduced radar signature through use of a special coating and refined aerodynamic shape. But there was little interest, as Russia in the 1990s just was not the place to be developing expensive new aircraft. One of the four airframes built is still at the Flight Research Institute, and was displayed at the bi-annual air show at Zhukovsky in 1993 and 1997.

Side view of the Yak-141 showing inlet door for lift engine.

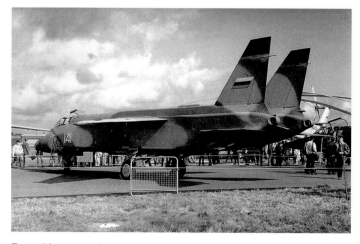

From this rear angle, note the nozzles used for yaw control mounted at the tip of each tail boom.

7
Lift Fan Thrust Augmentation VTOL Systems

The Lift Fan concept, as its name infers, produced lift by using a big fan located in each wing to blow air straight down. Only two such aircraft ever were built, although there were many conceptual designs. The Vanguard Omniplane was built and even tested in a wind tunnel, but as far as could be determined, never flown. The XV-5 demonstrated that the Lift Fan concept worked quite well.

In a strict sense, the Lift Fan is a type of thrust augmentation. As with any augmenter system, thrust was increased by converting the hot, high speed exhaust into a high mass flow at lower velocity and lower temperature. The fans were covered with louvers that resembled venetian blinds. They folded flush with the wing surface for conventional flight, opened for conventional flight, and could be modulated for quick thrust changes.

Lift Fans had many problems that never were overcome because of their limited development and flight testing. First was the fan size. In order to produce the ideal exhaust velocity, a very large fan was required. This resulted in a wing that was larger than the optimum size. Second, the fan thickness resulted in a relatively thick wing, limiting high speed performance. Third, the fan tended to produce a significant pitch up during the transition. Perhaps with further design and testing engineers might have overcome many of these problems.

VANGUARD 2C LIFT-FAN VTOL (U.S.A.)
System: Prototype built using many off-the-shelf components
Manufacturer: Vanguard Air & Marine Corp.
Sponser: Private company venture
Operating Period: 1950s time period
Mission: To develop an economical Lift-Fan VTOL system capable of performing both military and commercial missions.
The Story:
This is an amazing story since this Vanguard 2 was the private effort of a small two-dozen person company by the name of Vanguard Air & Marine Corporation. The company was directed by two former Piasecki engineers who brought considerable expertise to the undertaking.

Interestingly, the Model 2C attracted the interest of the Air Force's Wright Air Development Center, along with NASA's Ames Laboratory, which had been doing Lift-Fan wind tunnel research work.

Getting back to the Model 2C, there was the unbelievable price of under $40,000 each in a production run of a thousand units per year. It goes without saying that the plans for this model were ambitious to say the least!

A major influence in the design of the Model 2C was the utilization of technology from a previous VTOL Lift-Fan effort, the former Jacobs Convertiplane program. Vanguard had

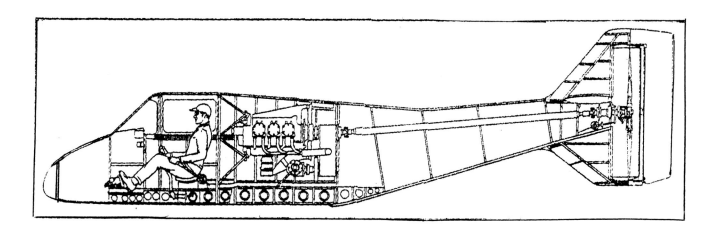

This internal shot shows the connection between the piston engine and the rear pusher prop.

The XV-5A was distinguished from the B version because of its US Army markings. (Ryan Photo)

actually purchased all the drawings, test data, and castings and patterns from that program, much of which was used in the Model 2C build-up.

The Model 2C was a somewhat conventional-appearing two-place low-wing configuration which mounted in mid-fuselage a 265 horsepower Lycoming 0540-AIA reciprocating engine. The company had hoped that the engine could later be equipped with a supercharger, therefore increasing the horsepower to 380.

Then comes the amazing coupling of the engine to the tail-mounted pusher propeller and the wingtip downward-pushing ducted rotors. The complicated set-up featured the engine power being funneled through a torsional coupling to a pulley shaft that turned the rear propeller. The ducted fans received their rotation by toothed rubber belting from a central transmission that connected to the rotor transmissions. An engaging clutch was in place between the central transmission and the pulley sheaves, an arrangement which permitted the disconnecting of the wing fans during conventional horizontal flight.

The wing rotors were three-bladed aluminum units whose bearings were capable of a thousand hours of operation. But since the ducted fans were used only a short time during each flight, Vanguard wasn't expecting any problems with them.

The Model 2C's wings featured a low aspect ratio and a thick airfoil having a modified NASA 4421 section. The wing consisted of two outboard panels which were bolted to a center section. The rotors and transmissions were supported within their respective ducts by four stators attached to the walls of the ducts.

In the normal mode of operation, the engine would be started with the wing rotors declutched and the tail propeller in low pitch. Then the wing closures would be opened and the clutch engaged. The vertical rise portion of the flight was then started and the tail propeller was thus increased for the

transition into horizontal flight. Then adequate forward velocity was reached, the duct closures were retracted, and the wing assumed its normal aerodynamic function with the wing rotors being declutched.

While the Model 2C was in a hovering mode, vehicle roll control was maintained by differential rotor pitch changes in the wing fans, while yaw and pitch control was accomplished by rudder and elevator movements. It was assessed that the tail rotor would provide pitch and yaw stability, even though it was a smaller unit that the tail rotors, since it possessed a long moment arm. Directional control for the Model 2C was provided by means of adjustable pedals and cables to the rudder.

Vanguard quickly noted that much greater performance than the basic Model 2C was possible with the substitution of a new 520 horsepower Lycoming ISO720 engine, which could provide a cruise speed of 200 miles per hour to a range of 550 miles. There was also a follow-on version to the 2C, known as the Model 8, which was to have a top speed of 260 miles per hour.

A military Model 7 version was to have the capability of carrying 32 troops or four tons of payload and a cruise speed of 288 miles per hour with a vertical climb rate of 300 feet per minute. Unfortunately, neither of these proposed later versions would make it to a prototype stage as did the Model 2C.

RYAN XV-5A AND XV-5B LIFT FAN VTOLS (U.S.A.)

System: Built-from-scratch research vehicle
Manufacturer: General Electric, Ryan Aeronautical Company
Sponsor:
Time Period: 1961 - 1974
Mission: Lift Fan research aircraft
The Story:
The Ryan XV-5 was another innovative VTOL aircraft that, in theory, should have solved all the problems associated with

vertical flight that continued to plague other research aircraft. It incorporated three powerful lift fans that were powered by the jet exhaust, to achieve vertical take-offs and landings.

The Army's Transportation Research Command issued a contract to build two XV-5As. General Electric, who developed the innovative engine/Lift Fan system, was the prime contractor. General Electric subcontracted with the Ryan Aeronautical Company to build the airframe. Although the XV-5A was to be a research aircraft, the basic size and configuration could be developed easily for other missions, such as utility, rescue, or even into a fighter aircraft.

General Electric began experimenting with the lift fan concept during the 1950s. These early developments included over 500 hours of propulsion system testing that culminated in a full scale lift fan test in the wind tunnel at NASA's Ames Research Center. As a result of this work, the Army awarded the XV-5A contract to General Electric in November 1961. The development of the integrated airframe/powerplant included over 600 hours of wind tunnel tests of scale models, full size models, and even the actual No. 1 XV-5A itself being tested in NASA Ames' giant 40 x 80 foot wind tunnel, the largest in the world. Testing the actual aircraft in the wind tunnel gave valuable insight into how to perform the transition by collecting data on stabilizer trim operation and fan louver positioning.

The XV-5A's basic configuration consisted of a five foot diameter lift fan buried in each wing and a three foot diameter fan in the nose for lift and pitch control. Exhaust diverted from the two J85-5B engines powered all three fans. The fans were really quite simple, having fixed blades, as compared to props whose blade angles could be varied. The blade tips were fixed to an outer metal ring. Additional turbine blades were attached to the outside of the ring. The fans were powered by diverting the engine exhaust past the outer turbine blades. The counter-rotating fans turned at only 2,640 revolutions per minute, comparable to propeller speeds. The design proved highly effective and saved weight by eliminating drive shafts and gearboxes. In addition, it improved reliability and simplified logistics support. The XV-5A had a useful load of 4,730 pounds. Being a research aircraft, most of the useful load consisted of data recording equipment.

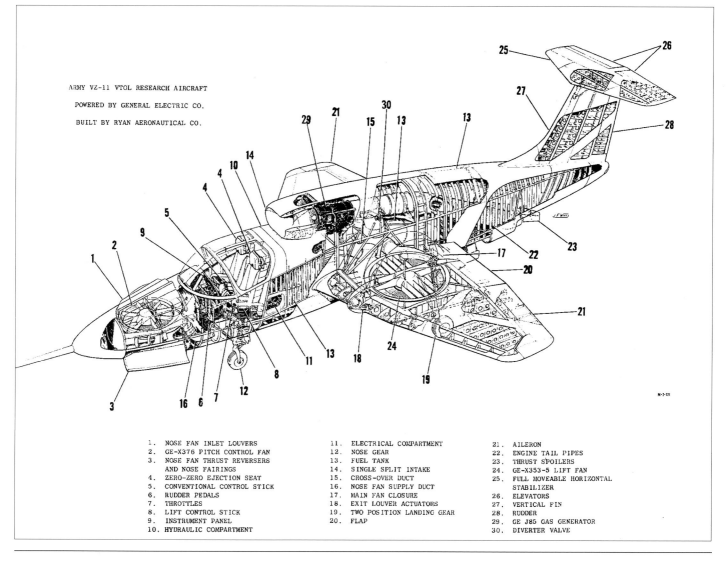

ARMY VZ-11 VTOL RESEARCH AIRCRAFT
POWERED BY GENERAL ELECTRIC CO.
BUILT BY RYAN AERONAUTICAL CO.

1. NOSE FAN INLET LOUVERS
2. GE-X376 PITCH CONTROL FAN
3. NOSE FAN THRUST REVERSERS AND NOSE FAIRINGS
4. ZERO-ZERO EJECTION SEAT
5. CONVENTIONAL CONTROL STICK
6. RUDDER PEDALS
7. THROTTLES
8. LIFT CONTROL STICK
9. INSTRUMENT PANEL
10. HYDRAULIC COMPARTMENT
11. ELECTRICAL COMPARTMENT
12. NOSE GEAR
13. FUEL TANK
14. SINGLE SPLIT INTAKE
15. CROSS-OVER DUCT
16. NOSE FAN SUPPLY DUCT
17. MAIN FAN CLOSURE
18. EXIT LOUVER ACTUATORS
19. TWO POSITION LANDING GEAR
20. FLAP
21. AILERON
22. ENGINE TAIL PIPES
23. THRUST SPOILERS
24. GE-X353-5 LIFT FAN
25. FULL MOVEABLE HORIZONTAL STABILIZER
26. ELEVATORS
27. VERTICAL FIN
28. RUDDER
29. GE J85 GAS GENERATOR
30. DIVERTER VALVE

An XV-5 is shown in vertical flight. (Ryan Photo)

The XV-5A had a design vertical take-off weight of 12,326 pounds. The two engines produced a total of 5,316 pounds of thrust, a reasonable thrust to weight ratio for normal, wing-borne flight. However, the thrust augmentation achieved with the three lift fans nearly tripled that thrust to 14,760 pounds, allowing the same engines to lift the XV-5A's total weight. Unlike other VTOL aircraft, there was no penalty for carrying either large engines sized for vertical take-off or lift engines that were dead weight throughout most of the flight. While this may sound like magic, it is simple physics. To explain it simply, the thrust is proportional to the product of the velocity and the mass flow rate of the exhaust. The power required to produce that thrust is proportional to the square of the velocity. Therefore, less power is needed to produce a given amount of thrust if the exhaust has a lower velocity. The three fans convert high speed jet exhaust into a high mass flow at low velocity, tripling the thrust! Other advantages were that the fuel used for a vertical take-off was comparable to the amount used for a conventional take-off, and the cool, relatively low speed fan exhaust permitted operation from almost any type of surface.

Of course, the Lift Fan concept had its own unique disadvantages. These included its volume and weight, and the difficulty in designing a suitable carry-through structure around it, especially in the wing.

Flight controls consisted of conventional-looking throttle, and a stick and rudder pedals connected to the elevator, rudder, and ailerons. The control surfaces became effective at about forty knots airspeed. For hover, a lift stick on the pilot's left side provided vertical control. The lift stick functioned like a helicopter's collective stick, but controlled the orientation of the fan louvers to modulate the fan thrust. At the top of the lift stick was a twist grip throttle that connected to the conventional throttles. The stick and rudder continued to control the pitch, bank, and yaw, but by modulating the fan louvers through the electronic flight control system. This mechaniza-

tion allowed the pilot to maintain his hand and foot positions throughout the transition into or out of VTOL mode. A stability augmentation system with limited pitch, roll, and yaw control improved the flight characteristics.

The two wing-mounted lift fans were covered with inlet doors on top and exit louvers on the bottom. During cruise, all doors and louvers were closed to produce smooth aerodynamic surfaces. During hover, the inlet doors were open and the exit louvers positioned parallel to each other. There were two ways to alter lift. Altering the engine thrust changed the lift fan speed. However, this produced a relatively slow response. For quicker response, the system positioned the louvers to restrict the flow, causing an instant thrust change. To begin the transition from hover to conventional flight, the louvers tilted aft to direct the thrust rearward. As the XV-5A accelerated past the stall speed, the pilot diverted the engine exhaust through the tail pipe and closed the doors and louvers. To transition from conventional flight to hover, the sequence was reversed. For simplicity, the flight control surfaces were always active, but obviously not effective at very low forward speeds when the vectored louvers provided all or most of the attitude control.

Thrust spoilers, installed just aft of the exhaust tail pipes, permitted the use of full engine power at reduced flying speeds. This allowed the fans to spin up to full speed in about one and a half seconds during transition and allowed the full transition without adjusting the throttles.

Crossover ducting between all fans and both engines insured that sixty percent of the total lift would still be available with only one engine operating. For normal landing weight of 8,000 pounds or less, the XV-5A could land with only one

One of the XV-5 prototypes demonstrating its rescue capabilities. (Ryan Photo)

All cleaned up, with its landing gear retracted, this XV-5 demonstrates its horizontal flight capabilities. (Ryan Photo)

engine operating. It could even take off on a single engine in conventional mode at 12,326 pounds gross weight. Vectoring the fan louvers aft permitted STOL take-offs. This would result in 8,400 pounds of forward thrust and 7,200 of lift thrust…all from engines producing a total of 5,316 pounds of thrust.

Following completion of the first XV-5A, Ryan performed tied down static tests at their factory at San Diego, at Edwards AFB, and at NASA Ames. These tests showed that most systems performed as expected. The one major problem uncovered was that the fan louvers were too weak to direct the fan exhaust at high power. Redesign of the louver actuation system caused enough delay to shift the first flight to a conventional, rather than a vertical, take-off. The XV-5A made its first flight on May 25, 1964, at Edwards AFB.

Test pilots from Republic Aviation, under contract to Ryan to perform initial test flying, performed ten very aggressive hours of flight testing. Using a variety of maneuvers, including stalls with flaps up and down, conventional flight between stall and 160 knots demonstrated that the aircraft met all requirements for stability, control sensitivity, and power requirements. Although they did not perform actual conversions to fan-supported flight, they opened the fan doors and louvers to the hover configuration as the XV-5A slowed to conversion speeds. Only a slight pitch trim change was needed to maintain control. Following the redesign of the louver doors, the first vertical take-off was made at Edwards AFB on July 16, 1964. Over thirty vertical take-offs were performed to test the hover mode. Hover proved to be simpler than in most helicopters. Fan-powered rolling take-offs and landings were performed at speeds just below the stall speed.

The second phase of flight testing demonstrated vertical take-offs, transition to conventional flight and back to a hover, and vertical landings. The first such full flight with a double transition was performed in December 1964. Early test results showed the XV-5A to be very sensitive to lateral wind gusts, primarily because of its relatively large side area. Pre-

cise altitude control with power was difficult because of the engine's slow response, but acceptable when using the fan louvers for thrust modulation. However, it still was poorer than in most helicopters. Severe buffeting was produced in hover below a five foot wheel height, but stopped by a wheel height of ten feet or greater. Prolonged hovering was limited to ten minutes due to structural heating.

Testing continued until one of the XV-5As crashed on April 27, 1965, killing the Ryan test pilot. Both XV-5As were performing a demonstration flight for several hundred reporters and other guests when one of the aircraft suddenly pitched downward from 800 feet and crashed into the dry lakebed. The cause was determined to be a premature conversion to the hover mode from conventional flight. The way the automatic sequence worked, as the louver doors opened, the computer gave a nose-down command to counter the doors' pitching moment. With the aircraft at a higher than normal

Reborn as the XV-5B, note the relocated landing gear and NASA paint scheme. (NASA Photo)

XV-5B in conventional flight. Note the closed louvers on the bottom of the fan. (NASA Photo)

speed for conversion, the doors produced a greater nose down motion than expected and the aircraft went out of control. The most likely cause was that the pilot hit the switch to initiate the conversion by accident, but there was no way to confirm this. To preclude this from happening again, the switch was moved to another location in the surviving XV-5A and it quickly resumed flying.

A year later, the XV-5A began another series of tests to evaluate its capability to perform search and rescue missions. The war in Vietnam was raging and was pushing helicopters to their limits of speed and range to perform dramatic rescues of downed airmen over the North. An operational air-craft based on the XV-5A design could provide a rescue air-craft that could enter enemy territory with the speed and maneuverability of a conventional jet, yet hover with acceptable downwash to perform the rescue.

Several modifications were made to the XV-5A in preparation for this test series. They were not so much for these tests, but based on experience gained from the flight tests to date. These included changing the roll sensitivity during hover to decrease pilot workload, simplifying the interconnect between the horizontal stabilizer and the thrust diverter (which provided automatic pitch compensation during transition) to improve reliability, ducting improvements to decrease inter-

Note the high mounted rear intake along with the nose doors being open of this XV-5. (NASA Photo)

nal drag, sequential conversion of the engines to allow each engine to be converted between lift and thrust modes separately (to allow improved acceleration and climb rates during conversion and more flexibility in short take-off operations), and improved insulation and cooling in the cockpit (which tended to get warm during hovering).

During one of these tests, on October 5, 1966, the XV-5A was involved in one of the most unusual crashes in aviation history, one in which the pilot was killed but the aircraft survived. Techniques were being developed for lifting a mannequin with a cable and sling. While hovering, the sling apparently became caught in the recirculating airflow and was sucked into the left lift fan. The loss in fan speed caused the aircraft to drop and roll to the left. The Air Force test pilot ejected, but was killed when his parachute failed to open. The fan then regained speed as the sling passed through, and the aircraft returned almost to level attitude by the time it hit the ground. The landing gear collapsed, but there was only minor, repairable, damage to the wing and fuselage. The crash certainly was survivable. The pilot made the correct decision for the situation he was in, but paid the ultimate price in a quirk of fate.

Up to this point, the two XV-5As completed 338 flights totaling 138 hours of flight time. In March of 1967, the Army transferred control of the surviving aircraft to NASA Ames Research Center in San Jose, CA. In preparation for NASA operation, the XV-5A was shipped to Ryan's facility at San Diego. Ryan repaired the damage caused by the crash and made several additional modifications to improve operations, maintenance, and reliability. These included moving the main landing gear outboard of the lift fans to improve ground stability and braking, installing improved braking, ejection, and fuel systems, and changing the cockpit layout and cockpit panel arrangements. All these changes added several hundred pounds to the empty weight. The aircraft was redesignated the XV-5B, and repainted in NASA markings: white with bright orange trim and bearing the NASA number of 705.

The reborn XV-5B made three checkout flights in the conventional mode on July 15 and 16, 1968, then flew to Edwards AFB on July 18. It continued on to Ames on August 28. NASA Ames operated the XV-5B until January 1974, performing research, such as terminal area guidance for V/STOL aircraft. Eventually, it was retired to the U.S. Army Aviation Museum at Ft. Rucker, Alabama.

8

Deflected Slipstream VTOL Systems

The Deflected Slipstream concept could well be considered as a variation of the Deflected Thrust technique. The only difference is, that instead of the engine thrust being deflected by close-in movable louvers in the rear of the engine, the deflection is caused by the thrust stream impinging on a control surface that is further downstream for the engine exhaust.

Deflected Slipstream lift-enhancement is also quite similar to a conventional prop aircraft carrying high-lift flaps. For the most part, the concept is not capable of producing true VTOL flight, but the advantages it does provide make it a viable concept when runway length is at a minimum.

The concept does not require the thrust to exceed the aircraft weight for short take-offs, as long as the combination of the wing and engine lift combined can support the aircraft weight at low forward speeds. A number of different concepts have attempted to combine the powered lift of the deflected slipstream with the aerodynamic lift of the aircraft wings.

Interestingly, it is possible to utilize the wing flaps in three different ways. First, the thrust can be blown underneath the wing directly onto the flap, thus causing the thrust to be turned downward and producing the vertical force to push the craft upward. Also, the thrust can be blown over the top of the wing and flaps (sometimes known as upper surface blowing) where the thrust is pulled down by the Condo effect. Finally, there is the so-called internal flow method where the flow comes from inside the wing before emitting and flowing over the flap.

The logical method of increasing the capability of the concept comes in two ways. First, the larger the area of the flap being covered by the thrust stream, the larger the lifting effect. And secondly, the angle of deflection of the flap will also directly affect the effectiveness of the concept. Theoretically, if the flap could be rotated to 90 degrees and the thrust was sufficient, a possible VTOL aircraft could be produced.

In addition to producing a short, near-vertical take-off and landing, these concepts also had the advantage of reducing take-off and landing distances, and landing speeds.

Interest in the deflection concept has been in place since World War I. Serious work began in the early 1950s when NACA started basic research on the concept. The advantage of this system was realized early because of its simplicity, since there are no mechanical or hydraulic devices required

to either pivot engines, or the even higher power requirements with tilting the entire wing.

But even with the simplicity factor in its favor, there are also a number of disadvantages inherent in the concept. As the angle of the thrust turning is increased, the thrust lost is increased correspondingly. It has also been shown that the thrust loss is greater closer to the ground.

Needless to say, a number of the Deflected Slipstream configurations that evolved presented some interesting appearances. On some, the trailing edge flaps were of the same size as the wings themselves. A number of the aircraft also featured wingtip fences to aid in containing the thrust upon the wing surface as much as possible.

The most famous of these STOL Deflected Slipstream aircraft developed were the Boeing YC-14 and McDonnell Douglas YC-15 testbed aircraft, which demonstrated both lower and upper surface blowing (see Chapter 14).

In 1979, NASA investigated the properties of a Deflected Slipstream VTOL fighter in its Ames Research Center 40x80 foot wind tunnel. (NASA Photo)

Two others of this STOL breed must also be mentioned, those being the NASA QSRA Upper-Surface-Blowing Quiet Short-Haul Research Aircraft. The model featured a four-turbojet transport with the engines mounted high on the upper wing. The engine exhaust was impinged directly onto the large trailing-edge flaps, thus providing a significant downforce, and in turn, an impressive short-take-off capability.

The French Breguet 940, which was first flown in 1958, was another example of a successful use of the Deflected Slipstream concept for transport applications. The transport used four 400 horsepower Turbomeca engines driving four interconnected propellers. The thrust stream covered most of the wing area and was turned down by the large full-span wing flaps. Like other Deflected Thrust aircraft, the roll control for the plane at low speeds was provided by differential propeller thrust, in this case, coming only from the outside propellers. Fully loaded, the 940 could take off and clear a 50-foot obstacle in about 600 feet, with only about 500 feet required to come to a stop after landing.

Another variation of this concept was utilized by Lockheed in an investigation of its C-130 Hercules military transport. The company investigated STOL improvements by application of a boundary layer control to the wing flaps and to all the control surfaces, thus increasing the effects of the Deflected Slipstream concept. Since the model did not have interconnected propellers, there could have been a large asymmetric off-balance of thrust in the event of the failure of an outboard engine. The concept was effective, though, with the plane demonstrating the capability to perform take-offs and landings over a 50-foot obstacle after only 1,500 feet.

During the late 1950s, the Fairchild Corporation investigated the Deflected Slipstream concept with its M-224-1 experimental model. The company indicated that it felt that a "Blown Flap" set-up would allow pure VTOL flight.

Testing the Ryan VZ-3RY. (Ryan Photo)

The M-224-1 sported a conventional high-wing design, an open cockpit, and a non-retractable tricycle landing gear. The model was one of the first aircraft to carry a T-type tail, although it carried substantial braces on each side. The wing was fitted with full-span leading edge flaps. Also, on its leading edge were mounted four airscrews, which were shaft driven by a single shaft-turbine engine installed above the fuselage. A small tail rotor provided control assistance during vertical and low-speed flight.

Following is a more detailed look at the most significant of the Deflected Slipstream VTOL concept.

RYAN 92 VZ-3RY VERTIPLANE DEFLECTED SLIPSTREAM VTOL (U.S.A.)

System: Built-from-scratch prototype
Manufacturer: Ryan Aeronautical Company
Sponsor: US Army, Office of Naval Research, NACA(NASA)
Operating Period: 1956-1961
Mission: Develop a Deflected Slipstream-style VTOL capable of fulfilling a reconnaissance and liaison mission profile, and able to operate from unimproved runways.
The Story:
It started off as a company program designated the Ryan 92, but would become a U.S. Army-sponsored program in 1956, given the designation of VZ-3. The contract, totaling $700,000, would be administered by the Office of Naval Research. Research was also conducted by the National Advisory Committee for Aeronautics (NACA) concerning the feasibility of the program.

Known as the Vertiplane, it was fabricated using simple construction techniques with a conventional high-wing design. However, the differences came with extensive double retractable wing flaps which extended far below and to the rear of the wing trailing edge. To visualize these flaps, you could say that they were the reverse of the upward-pointing winglets that are now a part of many modern transports.

The purpose of these interesting plates was entirely different, that being to contain the slipstream that was being funneled down the underside of the wing.

The contract was part of an Army VTOL research program, with Vertol also receiving a contract for construction of a different prototype version, but not using the Deflected Slipstream concept.

Ryan officials at the time felt that the design would be best suited to medium-velocity Army liaison, light passenger, and cargo transport roles.

Ryan certainly had heavy interest in VTOL development during this time period since it also had under development the Tail Sitter X-13 program.

To achieve that flow, the propellers were located on underwing pylons which had the slipstream from the twin props passing low under the 23-foot-span wing. The lift effect was also enhanced by large trailing edge flaps that could be curved downward, thus trapping the air between them and the wingtip plates.

As was the case with aircraft of the day, the VZ-3RY's propellers were wooden. (Ryan Photo)

Note the size of the trailing edge flaps in the fully-deployed position along with the spindly landing gear. (Ryan Photo)

Simply stated, this configuration achieves the same effect as a tilt-wing VTOL, except that the turning of the thrust is accomplished in a different manner. They have been called bucket VTOL craft because of this phenomena. Looking how the air is "contained," it's easy to understand how it could be so named.

The propellers carried the complete weight of the VZ-3 during lift before normal aerodynamic forces could come into play during transition and normal flight. Control was achieved during vertical lift by two different methods. First, roll control was achieved by differential propeller pitching, while both pitch and yaw control was achieved by vectoring engine exhaust.

Power to drive those significant propellers came from a mid-fuselage mounted thousand horsepower Lycoming T53-L-1 turboshaft engine. The engine was mounted inside the fuselage and drove two wing-mounted Hartzell three-blade wooden propellers.

During the take-off maneuver, the double, retractable flaps were fully extended, thus deflecting the engine slipstream practically straight down in a vertical direction. For transition to horizontal flight, the flaps were retracted as the craft gained forward speed, and the slipstream then flowed horizontally.

For the landing maneuver, the pilot made his approach with power on with some flap deflection. Next, he extended the flaps progressively and increased the power until the plane dropped in to touchdown at zero forward velocity.

The built-from-scratch fuselage, the nose of which seemed to be looking upward, was ungainly at best, and resembled a glider more than a powered aircraft. The diameter continued to increase from the tail to the back of the cockpit door. An external frame attached the enclosed cockpit's roof and screen.

The sheet metal-covered body was 26 feet long. Attached to the rear of the fuselage was a unique T-tail configuration.

Possibly in anticipation of hard landings, the landing gear struts were extremely thick and were attached to the bottom of the engine pilot, with two braces joined to the fuselage.

As was the case with many such intricate aircraft, there was wind tunnel testing accomplished by the NASA Ames Center. It was this testing that brought out the situation that a tricycle landing gear was required along with the addition of a small ventral fin for directional stability.

Ground testing would commence during 1958, with taxiing testing being the first undertaking. But it wouldn't be until January of 1959 before the first flight would take place at Moffett Field, California. An accident shut down the flight test program for a short time, but the craft was back in the air the following month.

To the dismay of company engineers, it was determined that the engine was not up to their design mission in that it required the plane flying into a headwind in order to maintain a hover.

The Army then handed over the prototype to NASA for further testing following 21 successful flights. The transition was far from uneventful, though, as the plane was involved in an unplanned maneuver, causing the pilot to eject and severely damaging the craft.

Side view of deployed rear flaps. (Ryan Photo)

Rear view of VZ-3RY with rear flaps deployed. (Ryan Photo)

Sitting on all its landing gears, the VZ-5FA looks a little more like a normal aircraft. (Fairchild Photo)

After consultation, it was decided to rebuild the VZ-3, but a number of modifications were added. The new open cockpit included an open cockpit and a fabric nose section. The flying program would continue the following year, investigating a number of low-speed characteristics.

FAIRCHILD 224 VZ-5 FLEDGLING DEFLECTED SLIP-STREAM VTOL (U.S.A.)

System: Built-from-scratch prototype
Manufacturer: Fairchild Corporation
Sponsor: US Army
Operating Period: Late 1950s time period
Mission: To design and develop a prototype four-engine Deflected Slipstream small transport for the U.S. Army
The Story:
The Fairchild 224 VZ-5 was actually designed from the beginning to have the same capabilities as the VZ-3. But in the case of the VZ-5, the story is a short one since this unique one-of-a-kind aircraft would never fly in free flight.

For a craft so small, a 34-foot-long fuselage and a 33-foot-span wing, the system still carried four wing-mounted turboshaft engines. In this application, powerful General Electric YT58-GE-2 engines, each capable of 1,024 horsepower, drove three-bladed Harzell propellers.

The pilot resided in a small open cockpit in the nose, but there was also room for an observer or monitoring equipment in a seat location behind the pilot. The location of that cockpit was interesting in that it was located extremely low when compared to the position where the wing joined the fuselage. It was as though the cockpit was removed from the rest of the aircraft.

The aircraft had the capability to either rest on its forward tricycle landing gear, or sit on its two main wheels in addition to two main wheels and a fixed tail skid. In that latter orientation, the Fledgling was effectively starting off its upward flight with a 30-degree angle-of-attack. This situation effectively enhanced the bucket effect that was achieved by the engines directly thrusting into the large trailing edge flaps, thus producing considerable near vertical downward thrust. With that nose-high attitude, it appeared that the plane was already in flight even though it was still sitting on the ground.

Vertical tabs on the end of each wing also aided in the containment of that downward thrust.

As with any aircraft of this type, additional control was required. With this aircraft, pitch control was accomplished by small rotors that were located atop the VZ-5's T-tail.

Hopes were somewhat high for the possible future of the plane, but the test program would not venture beyond tethered flights, which were conducted late in 1959.

Even with the model sitting in its ground configuration, the VZ-5FA looked like it was already taking off. (Fairchild Photo)

Note the wing end caps on the X-1 which helped contain the air flow of the Deflected Slipstream concept. (Robertson Photo)

KAMAN K-16 DEFLECTED SLIPSTREAM VTOL SYSTEM (U.S.A.)

System: Built-from-scratch prototype
Manufacturer: Kaman Corporation
Sponsor: US Navy
Operating Period: Unknown
Mission: To develop a Deflected Slipstream research aircraft for U.S. Navy applications
The Story:
The Kaman was another of numerous Deflected Slipstream prototypes to explore the concept. The K-16 system is unique, with a large full-span extendible wing flap to deflect the rotor slipstream downward, thus providing lift. The system also had the provision for increasing the angle of incidence of the wing with respect to the fuselage. That allows the fuselage to remain in a horizontal attitude in both hovering and low-velocity flight conditions.

So the K-16 would actually have to be classified a dual mode VTOL system with both the obvious Deflected Slipstream concept, along with a partial tilt-wing capability. However, since the wing had only a 50 degree tilt capability, the model is classified as a Slipstream Deflection type since a majority of the vertical lift comes from the deflecting flaps.

The K-16 was a relatively small aircraft, weighing only four and one-half tons. Power came from a pair of wing-mounted T-58 gas turbine engines which were connected to 15-foot diameter rotors interconnected with cross shafting.

Pitch control for the craft was accomplished with a technique similar to that of other Deflected Thrust systems, that being the use of a helicopter-type cyclic-pitch control system. Kaman felt that this system would eliminate the complicated tail rotor set-up of many such systems.

Wind tunnel testing was a prominent part of this program, with the model being tested at the 40x80-foot wind tunnel at the NASA Ames Research Center. It was followed by the flight test program.

ROBERTSON X-1 DEFLECTED SLIPSTREAM VTOL (U.S.A.)

System: Built-from-scratch prototype
Manufacturer: Robertson Aircraft Company
Sponsor: Company program, later US Army involvement
Operating Period: 1956-1957 time period
Mission: To develop a twin-engine Deflected Slipstream transport.
The Story:
The initial importance of this Deflected Slipstream VTOL program is emphasized from the fact that the Robertson Aircraft Company was actually formed to build such a system.

The Robertson system was similar to several other such systems that were built during this period where interest was high on the Deflected Slipstream concept. The program was initiated in 1956, but would be short-lived, with no free-flight tests attempted with the lone prototype.

Power was significant on the craft, with a pair of Lycoming GSO-480 supercharged reciprocating 340-horsepower engines which were mounted on vertical underwing pylons. The engines drove a pair of 106-inch three-bladed Hartzell propellers.

All control for the craft was provided by a sliding flap system with a double-slotted full span trailing edge flap. There were no control augmentation devices such as used on the

The X-1's trailing edge flaps almost touched the ground when in the deployed position. (Robertson Photo)

other Deflected Slipstream programs. The flaps were of an amazing size, actually larger than the wing itself, when they were deployed. In the deployed position, the flaps almost touched the ground. They gave the look that a curtain had been drawn behind the plane.

The wings were extremely short compared to conventional aircraft of the same size. Joined high on the fuselage directly over the cockpit, the wing was fully loaded with engines and wingtip tanks. The fuselage was of a standard configuration, with a standard tail consisting of a vertical fin and positive dihedral horizontal surfaces.

The Robertson design in its flight mode called for the full retraction of all flaps once the plane had achieved conventional horizontal flight. An interesting innovation on the plane was the location of the pair of wing-tip fuel tanks which were attached to vertical flaps and aided in the function of air con-tainment, thus increasing the efficiency of the lifting phenomena. Where the flap attachment points joined the wingtips, the wingtips were completely square, adding to the "short-wing" look of the model.

The Robertson machine carried a tricycle landing gear which had to support the aircraft at a significant height because of the low-hanging engines, with their large propellers.

The test program was brief, with a tethered flight made in January 1957, but for some reason, the program was apparently not pursued after that time.

The X-1 was just the first version of the X-1, the second of which would never take place. Never-the-less, though, there were plans for replacing the reciprocating engines with Lycoming T-53 turboprops. To take advantage of the increased power from the jet slipstream, it was planned that stainless steel would be added to the wing structure.

9

Ejector Augmentation Thrust VTOL Systems

The ejector was a simple enough concept for increasing a jet engine's thrust. It was simply a chamber open at each end into which the jet exhaust exited. The hot, high speed exhaust drew in ambient air, which increased the overall thrust. As simple a device as it was, achieving the proper shape was critical to making it work, an accomplishment only partly achieved by designers.

Only two aircraft were built and tested that used the Ejector Thrust Augmentation concept. These were the XV-4A and the XFV-12A. The XV-4A was a research aircraft and achieved a degree of success. However, the augmenter occupied most of the fuselage volume, severely limiting any mission that would require carrying anything internally.

Perhaps based on the XV-4A's success, the Navy felt the Ejector Thrust Augmentation concept was mature enough for a production aircraft and jumped into developing the XFV-12A. It was a prototype for a new fighter, not a research aircraft. Rather than using valuable space in the fuselage for the augmenter, panels on the top and bottom of the wing and canard opened to form the augmentation chamber. This was a great idea, except that the results of theoretical predictions and laboratory experiments far exceeded the results of the real hardware. The XFV-12A never lifted off the ground.

LOCKHEED XV-4A HUMMINGBIRD AUGMENTED THRUST VTOL (U.S.A.)
System: Built-from-scratch research aircraft
Manufacturer: Lockheed-Georgia
Sponsor: U.S. Army
Time Period: 1959 - 1964
Mission: Augmented Thrust research aircraft
The Story:
The XV-4A was another research aircraft built to test the thrust augmentation concept in which more lift is produced for vertical flight than the aircraft's jet engines produce in thrust. It was developed by the Lockheed-Georgia Company under a contract from the U.S. Army. Lockheed proposed their Model 330 Hummingbird in August of 1959 as an integrated VTOL aircraft system with electronics designed for battlefield reconnaissance and target identification. The Army chose to proceed only with the design, manufacturing, and flight testing of two research versions of the aircraft, leaving off the reconnaissance systems. Initially designated the VZ-10, but

later changed to XV-4A, the program was established in June 1961. Only 53 weeks elapsed between signing the contract and first flight in conventional mode.

The primary mission was to determine the feasibility of the augmented jet ejector concept for use in VTOL operations, investigate handling quality requirements for VTOL aircraft, and to perform a semioperational evaluation of the aircraft. It was hoped that test results would provide the basis for a future manned VTOL reconnaissance aircraft. Lockheed built two test aircraft.

The XV-4A featured a mid-fuselage mounted wing, T-tail, tricycle retractable landing gear, and side-by-side seating (although the right seat was occupied by recording and other equipment). Two Pratt and Whitney JT-12A-3(LH) turbojet engines provided all lift and cruise power. They were installed in pods on either side of the fuselage at the upper side of the wing root. During conventional flight, thrust from the engines flowed rearward and exited through conventional nozzles. During vertical flight, the jet exhaust was diverted into a jet-ejector vertical lift system located in the fuselage. In theory, the jet augmenters would increase the total engine thrust of 6,600 pounds to a level well above the aircraft weight.

The XV-4A was designed to be a tough little airplane, with a wing span measuring only 25 feet. G-limits were +6 and -3. It could withstand gusts of 66 feet per second at the maximum design cruise speed of 350 knots and gusts of 50

XV-4A in conventional flight with augmentor doors closed. (Lockheed Photo)

feet per second at the maximum dive speed of 450 knots at sea level. However, the maximum speed demonstrated during the flight test was only 170 knots. The landing gear was designed for a sink rate of 16.6 feet per second during VTOL landings and 10 feet per second during conventional landings. Fourteen feet per second was the most actually demonstrated in a VTOL landing. Rocket-powered ejection seats from a Douglas A-4E were modified for zero-zero performance.

The jet-ejector vertical lift system consisted of two manifolds, with each manifold having ten sets of downward pointing nozzles. The engine exhaust was split such that half went into five nozzles in each manifold, avoiding asymmetric lift in the event of an engine failure. The manifolds were open at the top and bottom, so that ambient air could be pulled in. The chambers were made of titanium-alloy skin backed by steel stiffeners. In the process of drawing in outside air, the overall mass flow rate increased by about forty percent. This resulted in a significant velocity decrease and overall thrust increase. The end result was that the jet ejectors produced more thrust than the engines alone could produce. The manifold chambers slanted aft, so that the XV-4A lifted off and hovered in a 12 degree nose up attitude, which aided the transition to and from conventional flight. The nose gear could be shortened to give a better attitude for conventional take-offs and landings. Long, narrow doors covered the top and bottom of the fuselage. The doors could be positioned to help modulate the amount of secondary air, and thus the thrust produced. The temperature of the mixed flow leaving the augmenter was about 300°F, compared to about 1200°F for the exhaust leaving the engines. By virtue of the low temperature and relatively low velocity, a person could stand within a few yards of the hovering XV-4A and not get blown over.

The wing was equipped with a single-slotted trailing edge flap and ailerons. Two flap positions could be selected, up or down. All fuel was carried in the fuselage between the augmenter chambers. A blowing boundary layer control system was used on the horizontal stabilizer during vertical flight to improve tail effectiveness.

Empty weight was 4,995 pounds, and vertical take-off weight was 7,200 pounds. Being a research aircraft, weight reduction was not the highest of priorities. A production version certainly would have had a lower empty weight. Engineers predicted that the augmenter design would yield a vertical take-off thrust of 8,375 pounds. Lift augmentation depends on a number of geometric factors. The principal one is the ratio of the outlet area to the nozzle area. The larger the ratio, the greater the augmentation. The XV-4A was designed with a ratio of 13.6, which was expected to produce an augmentation of 1.4. The maximum vertical thrust actually attained in flight was only 7,800 pounds. Given that the engines lost about ten percent of their thrust in diverting the exhaust to the augmenters, the net augmentation was just over 1.3. Successive ejector nozzle designs were tried, but none ever produced the 8,375 pounds of lift force that was hoped for.

Aircraft control during conventional flight was by movement of the elevator, ailerons, and rudder, which were connected hydromechanically to the control stick and rudder pedals. Mechanical connections to all surfaces were boosted hydraulically. During hover, pitch control was maintained using a pair of downward-pointing pitch thrusters located at the nose and the tail. They were rotated differentially for yaw control. These nozzles produced about 450 pounds of constant thrust. For redundancy, one engine controlled the nose set, and the other engine controlled those at the tail. These thrusters were powered by engine exhaust. Roll was controlled using two reaction control thrusters located at each wing tip, powered by engine compressor bleed air. For simplicity, all thrusters and control surfaces operated at all times, eliminating the need for complex switchover algorithms as the aircraft transitioned between flight modes.

A limited authority stability augmentation system improved flight characteristics. In flight, the XV-4A was unstable in pitch and roll without the stability augmentation turned on. Unaugmented, the pitch motion was divergent, with no tendency to oscillate. The roll motion primarily was an interchange between bank and sideslip, gradually diverging. Yaw motions were stable without augmentation. The stability augmentation system was relatively simple, being adapted from a helicopter. While considered adequate for an experimental aircraft, it was never intended to be representative of a production system, and could be overridden by the pilot using large stick or pedal inputs.

During transition from vertical to conventional flight mode, the elevator automatically moved 30 degrees trailing edge down to compensate for a nose up pitch moment caused by the augmenters and doors. To prevent the tail from stalling at these large angles, the boundary layer control system blew high-speed air over the upper leading edges of the horizontal stabilizer and elevators during VTOL fight. No special features were required to aid the roll and yaw control systems during transition, other than the stability augmentation already in place.

The flight test program was conducted at Dobbins Air Force Base, near Marietta, GA. Both aircraft were rolled out in spring 1962, separated by a few months. The first conventional flight was made in July 1962. This flight revealed a number of small deficiencies, such as an engine thrust misalignment and excessive trim changes with flap and landing gear retraction. Eight conventional flights by four different pilots demonstrated the overall good flight characteristics up to 250 knots and 10,000 feet of altitude. Then followed a series of hovering flights in a test rig, well out of ground effect, then tethered hover flights. The first tethered hover was on November 30, and the first transition performed on February 11, 1963.

Because of worse than expected augmenter performance, the aircraft was usually flown with very low fuel loads to decrease weight. Takeoffs required full power just to get into the air, making precise height control right after take off impossible until a quantity of fuel was consumed. Hover landings were accomplished by reducing power and descending, then increasing power to arrest the descent just before touchdown.

During hover, the XV-4A could move forward, backward, and sideways by making very small pitch or bank inputs. By keeping the inputs small, changes in the altitude were insignificant, but pitch was much more sensitive than bank. Lateral motions were reported to be solid and easy to make. Similarly, yaw motions were not a problem. Throttle movements controlled height. Overall, hover control was considered easy.

When the wheels were less than three feet above the runway, the augmenter exhaust reflected off the surface and hit the bottom of the wing and fuselage. This produced a positive fountain effect, but also caused a high frequency vibration. The vibration virtually stopped when the wheel height was greater than three feet, or when the aircraft was hovering in a forward direction directly into the wind. The reflected exhaust obviously was missing the aircraft altogether. In a prolonged, stable hover at full power near the ground, the warm exhaust that reflected off the runway also traveled around the aircraft and was ingested into the engine inlets, producing a measurable thrust decrease.

Translation from hover to forward motion was performed in several steps. First, the nose was lowered slightly to provide a forward thrust component. Because of the limited vertical thrust, the aircraft usually settled back onto the runway once or twice as it accelerated. After reaching about 20 or 30 knots, acceleration improved because thrust increased as reingestion of hot gasses stopped. As the wings began generating lift, the nose could be lowered more to provide more horizontal thrust. At this point, it accelerated and climbed fairly well. At about 80 knots, the number one engine was switched from lift mode to thrust mode. After this, the best acceleration was achieved by *decreasing* the thrust on the number two engine! This was because the augmenters produced drag as the aircraft moved forward, forcing the airflow to turn almost 90 degrees as it entered the augmenter chambers. Decreasing the augmenter thrust decreased the forward drag. The pilot completed the transition to forward flight at about 120 knots as the number two engine was converted to the thrust mode and the doors on the top and bottom of the fuselage were closed. A slight improvement in the lateral directional stability was noted as the doors closed. The transition from conventional flight to hover was accomplished in essentially the reverse order, except that the engines were retarded to idle and placed in the lift mode to slow to hovering speed.

Conventional take-offs required only 1,600 feet of runway at a gross weight of 7,200 pounds. Conventional landings took advantage of the drag caused by the augmenters to reduce the landing roll. After touchdown, the ejector doors

were opened and both engines were placed in the lifting mode. This technique produced a landing roll of 3,500 to 4,000 feet. A drag parachute was installed, but never used during flight tests because this technique was so effective.

During up and away flight, abrupt pull-ups of up to 1.8g were made. Upon stick release, the response was deadbeat, with no oscillation. Rudder kicks and sideslip releases resulted in two lateral-directional oscillation cycles that then abruptly stopped with the left wing down 25 to 30 degrees. The stability augmentation system was never fixed to eliminate this tendency because optimizing the flight characteristics was not one of the program objectives.

The XV-4A made 82 hovering flights for a total of eight hours flight time. It also made 69 conventional flights totaling 28 hours. Seven complete transitions from hover to conventional flight and back to hover were made. In-flight engine shutdowns and restarts were accomplished easily. Cruise flight on a single engine also was reported to be easy. The XV-4A had excellent acceleration and deceleration in response to throttle changes. Sea level rate of climb was 18,000 feet per minute on both engines, and 5,500 feet per minute on one engine. It could reach 10,000 feet in 58 seconds. Stalls were quick and without buffet (which is typical of a high tail that is out of the wing slipstream), and were accompanied by a rapid left roll and downward pitch. However, recovery was easy, requiring only the application of power and opposite aileron.

One of the XV-4As crashed on June 10, 1964, killing the civilian Army test pilot and destroying the aircraft. According to the pilot in a chase plane, the XV-4A had just completed a transition to vertical flight at 3,000 feet when it went out of control.

Upon completion of the flight test program, efforts were made to improve the marginal augmenter performance. Numerous new manifold and ejector configurations were tested with the second aircraft mounted in a static test rig. The best that any of these configurations produced was 8,640 pounds of vertical thrust (which included 450 pounds from the pitch control nozzles), still well below the original hoped-for value. Other test configurations made of heavier materials gave some better results, but a flight-qualified version made of

With augmentor doors open, this XV-4A takes to the air. (Lockheed Photo)

lighter materials could not be made for testing in the aircraft before the program terminated in 1964.

NORTH AMERICAN ROCKWELL XFV-12A AUGMENTED THRUST VTOL (U.S.A.)

System: Built-from-scratch prototype
Manufacturer: North American Rockwell
Sponsor: U.S. Navy
Time Period: 1971 - 1981
Mission: Fighter attack aircraft for fleet defense
The Story:

The unconventional-looking XFV-12A, built by North American Rockwell at its Columbus, Ohio, facility, was designed to be a supersonic, single place, fighter attack aircraft. There was great optimism that combining the technology of thrust augmentation with a canard wing would eliminate most of the limitations of other VTOL aircraft. An optimal design was expected that would have the right amount of thrust needed for vertical take-offs and landings and an aerodynamic configuration good for both supersonic flight and hovering. The XFV-12A used a single Pratt & Whitney F401-400 turbofan engine, with afterburner, and a special nozzle that exhausted the thrust out the back for conventional flight and diverted it to the wing and canard-mounted thrust ejectors for vertical flight.

The innovative concept behind the XFV-12A was the Thrust Augmenter Wing. This design took engine exhaust, diverted it through internal ducting to the wing and canard, and then exhausted it downward through ejector augmenters. At this point, the hot exhaust pulled cool ambient air into the flow, slowing the velocity, increasing the mass flow, and increasing the thrust. Initial research showed that engine thrust could be doubled using this thrust augmentation concept. This may sound like getting something for nothing, but the laws of physics always apply! The jet exhaust has a certain amount of energy, but the thrust is determined by factors such as exhaust temperature, mass flow rate, and exit velocity. Heat is a form of energy, and the heat in the jet exhaust is wasted energy. By sucking in large quantities of cooler air and carefully mixing it with the hot exhaust (7.5 pounds of ambient air for each pound of exhaust exiting the nozzle), the heat is exchanged for greater mass flow. Thus, the energy contained in the jet exhaust is used more efficiently to produce more thrust. Lift thrust of 21,800 pounds was expected from an engine that produced 14,070 pounds thrust for conventional flight. Design vertical take-off weight was 19,500 pounds, but with a 300 foot rolling take-off, an additional 5,000 pounds of payload could be carried.

The XFV-12A design promised other advantages. Problems inherent in other VTOL aircraft, such as surface erosion and reingestion of hot exhaust gas and debris, would be minimized because the augmenter exhaust would have a lower speed (about 400 feet per second, as compared to 2000 feet per second for the engine exhaust) and lower temperature than the jet exhaust itself. It also would provide adequate thrust for vertical operations without paying the penalty of carrying a large engine sized for vertical flight and/or lift engines that were dead weight throughout most of the mission. The vertical stabilizer was replaced by wing endplates that served a dual purpose. First, they would provide good directional stability at all angles of attack because they always would be in smooth air that was not disturbed by the fuselage. Also, their location at the wing tip was further aft than a vertical tail could have been located on the fuselage, allowing them to produce more yawing moment. Second, they would improve the wing's efficiency by breaking up most of the wing tip vortex, in effect, making the wing perform as a longer, more slender one. Even the positioning of the wings, low for the canard and high for the main wing was significant; during transition, most of the free stream air reaching the main wing was undisturbed by the flow around the canard.

The history of the XFV-12A originates in about 1970 with a Navy initiative called the Surface Control Ship. The Navy was concerned with the high price of aircraft carriers. Surface Control Ships would be very small, relatively inexpensive aircraft carriers. They would be little more than floating

XFV-12A technical prototype VTOL. (North American Aviation Photo)

Artist's concept of XFV-12A VTOL in flight. (North American Aviation Drawing)

launch pads, having minimal capability. Their size would permit only vertical take-off and landing aircraft that would perform the entire ship mission. For example, the ship would have no defensive capability; defense would rest solely with the aircraft they carried.

There also was concern for the rising price of sophisticated Navy fighters, such as the F-14 that was just entering flight testing. Part of the high cost was from the length of time it took to study a requirement and perform a precise design. For the XFV-12A, it was decided to try a prototype approach to reduce lengthy development times. This differed in that a component would be built and tested following a much simpler design and analysis process than usual. Improvements would be made as the hardware was tested and flaws in the design were discovered. Also, to help decrease constraints, the Navy specified only the general purpose and characteristics of the aircraft. Industry would be free to tell the Navy what was needed to meet those requirements.

To cut procurement time, the Navy used a highly streamlined procurement process. A Letter of Interest was released in November 1971. It contained no specific requirements or specifications, normal for such a document, but encouraged innovative ideas using advanced technology. Two different aircraft actually were specified for operation from the Surface Control Ship: a high performance short range fighter attack aircraft and a long range sensor carrier with weapons delivery capability. Proposals would be limited to 20 pages and due within 45 days. Contract awarded to the winning proposals would follow in another 45 days. Being ready for such a Navy requirement, North American already had invested heavily in the concept, performing much in-house research over the previous two years.

By early January 1972, 11 proposals were received for the fighter attack aircraft and 19 for the sensor carrier. While all of the fighter proposals were for VTOL jets, the sensor

carrier proposals included jets, turboprops, tilt wings, tilt props, and helicopters. The Navy selected the NR-356 proposed by North American Rockwell for the fighter attack aircraft, but dropped any further plans for the sensor carrier.

North American chose to perform the aircraft development at its Columbus, Ohio, facility for two reasons. First, they were near the Air Force's Aeronautical Research Laboratory at Wright-Patterson AFB in nearby Dayton. This organization, which eventually became Wright Laboratory and then the Air Force Research Laboratory in the 1990s, was performing much of the research in ejector augmenter technology. Second was the business situation. The last RA-5C Vigilante reconnaissance aircraft was about to be delivered, and the division's other two products, the T-2 Buckeye trainer and the OV-10 Bronco observation/utility aircraft, would continue at only very low production rates. This translated to a large pool of experienced engineers and extensive production facilities being available and at risk of being lost if not used.

The basic concept of the Thrust Augmenter Wing appeared sound, being based on research using hardware that already had been built and tested. A set of three flaps, two opening downward and one upward, formed channels on the wing and canard. A series of nozzles, or ejectors, were placed spanwise along the base of the center flap. When the flaps were extended, a large space was opened to allow ambient air to join the jet exhaust flow. For wing-borne flight, the flaps folded in to form an integral part of the wing surface. For attitude control during vertical flight, the two downward opening flaps could move in and out to modulate the amount of ambient air sucked into the flow, and thus change the thrust produced by each wing without changing the engine speed. The four ejector assemblies formed a basic "four post" configuration. They could be modulated for attitude control, eliminating the need for thrusters at the wingtips, nose, and tail.

Artist's concept of XFV-12A VTOL shown in carrier operations. (North American Aviation Drawing)

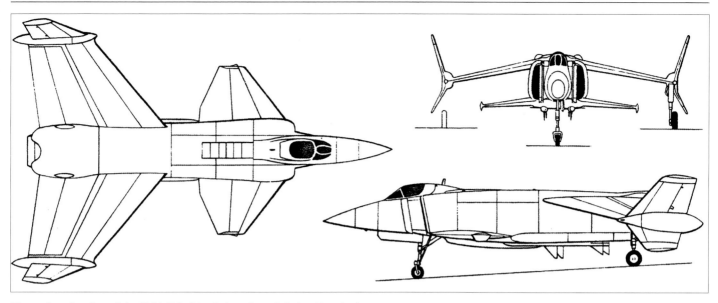

Three-view drawing of the XFV-12A. (North American Aviation Drawing)

The critical measurement of how much thrust augmentation could be produced is the ratio of thrust produced by the augmented wing to the amount of thrust delivered by the ejector nozzles without any augmentation. This is called the augmentation ratio. Laboratory tests showed that augmentation ratios greater than 2.0 were possible on very long flaps, but probably would be difficult to achieve on an actual aircraft. One of the few specifications in the contract was that an overall augmentation ratio of 1.55 was to be demonstrated within 11 months of contract start. Continued development would be based on meeting this milestone.

The use of fly-by-wire flight controls was rejected in favor of the more conventional mechanical linkages. Rockwell decided that this technology was too new and posed too much technical, cost, and schedule risk. To further save time and development risk, the cockpit and forward fuselage from an A-4 and the landing gear, inlets, and wing box from an F-4 were used.

The thrust diverter was a simple, yet innovative design developed by Pratt & Whitney. The nozzle contained a plug that could slide fore and aft. As it slid back, it sealed the nozzle opening. Simultaneously, diverter doors on the side of the engine opened to allow the exhaust to be directed into ducts that led to the wings and the ejector nozzles. The ratio of the nozzle and diverter openings was matched carefully to allow constant flow through the engine throughout the transition. Five ball screw actuators moved the sliding plug and diverter doors. To save costs, the hardware was not quite production quality, but of a sound enough design to represent a future production system.

Despite the plan for a quick award, the contract for design, construction, and testing of the prototype was not signed until March 1973. The first laboratory test of a 0.2 scale model of the wing augmenter showed an augmentation ratio of 1.4,

giving everyone optimism that the required value of 1.55 was achievable. Another difficult area that required careful design was ducting for the hot engine exhaust. It required minimal leakage while accounting for thermal expansion and contraction, pressure losses due to turns, and in-flight flexing. The final design was made from titanium and met all requirements.

While other systems were coming along nicely, the augmentation ratio continued to be a problem. In order to work, the design required an augmentation ratio of 1.6 for the wing and 1.5 for the canard. The goal for the canard was proving much more difficult because of its smaller size. By 1975, North American conducted tests on full size wing and canard ejectors, but only achieved augmentation ratios of 1.49 for the wing and 1.29 for the canard. However, the Navy continued to fund the development because the goals still seemed achievable.

The single prototype finally was completed and rolled out in May 1977. Although the original contract called for two prototypes to be built, one had been eliminated by this time in order to cut escalating costs. Throughout the design and fabrication process, the weight grew from the original estimate of 13,800 pounds to 15,596 pounds, of which 300 pounds were due to increases in the engine weight. Static thrust measurements began in July. Numerous problems, not unusual in testing a prototype aircraft, arose, such as the failure of titanium vanes in the augmenter ducting, cracks in the diverter area, and the deformation of ducts. These problems were fixed, but augmenter performance was difficult to determine because suckdown and fountain effects could not be separated out from the overall aircraft performance. Methods to suspend the XFV-12A in the air, to get it out of ground effect, were proposed so that the augmenter performance could be studied better. It finally was decided to transfer the aircraft to NASA's Langley Research Center in Hampton, Virginia, in order to suspend it from the old lunar lander test rig.

Proposals were made to fly the XFV-12A to Langley in its conventional flight mode. This was rejected for several reasons. It would first require extensive flight testing out of Rockwell's Columbus facility. This was considered far too risky for such an unconventional aircraft, as only one was built. Also, funds would not be provided because the first big milestone, demonstrating the required augmentation ratio, had yet to be demonstrated. Thus, it was decided to transport the XFV-12A using an Aero Spacelines Super Guppy. This was done in November 1977.

Suspended from the test rig at NASA Langley, the augmentation ratios proved to be dismal. Even after extensive diagnosis and numerous modifications, the best ever achieved was 1.19 for the wing and 1.06 for the canard. The Navy and Rockwell conceded that the hoped for augmentation ratios demonstrated in the laboratory could not be obtained in a real aircraft due to geometric differences with laboratory test units and other real-world factors that do not exist in the laboratory. Also, although hot gas reingestion was significantly less than for other VTOL aircraft, it still existed, causing a further decrease in the engine thrust.

With the planned purchase of Harriers for the Marines, interest in VTOL aircraft continued within the Navy. Limited funding was provided for another three years to continue studying the augmenter performance. The augmenters from one wing and one canard were removed and shipped to Rockwell's Columbus facility for further study. There were numerous advocates for continuing thrust augmentation research and using the XFV-12A as a research aircraft, but significant funding never was obtained. Unfortunately, the aircraft never even flew in the conventional mode to assess its flight characteristics or the performance of the canard design. The original contract for $46.9 million had stretched to about $97 million over ten years, and that was enough. The Navy finally terminated the program in 1981.

The XFV-12 was disassembled at Langley and shipped to NASA's Lewis Research Center in 1984 for unspecified possible future use. It was placed in storage at Lewis's Plum Brook facility near Sandusky, OH. In early 1998 it was scrapped as NASA could find no use for it and could not find any museum that wanted it. This time, only the cockpit was saved for possible future use.

10
Tail-Sitter VTOL Systems

The true VTOL would have to be considered a vehicle which ascends vertically while in the vertical attitude. Most vertically-lifting VTOL vehicles rise with the vehicle in the horizontal position with pivotable engines, tilt-wings, deflected thrust, or other techniques used to acquire the downward thrust.

With the Tail-Sitter concept, the vehicle begins its flight from a vertical attitude and rises vertically before initiating a transition to horizontal flight. It's actually probably the purist of the VTOL systems, although the concept always had a payload limitation.

The nature of the Tail Sitter concept dictates that the vehicle be short and squatty in shape, which negates the problem of a higher center of gravity and a greater tendency to tip over during the lower thrust-to-weight ratios of early vertical flight. That shaping was definitely evident in the 1950s Tail Sitter concepts.

One advantage of the technique is that the line of thrust is along the centerline of the fuselage, which could have aided in the stability of the vehicle. Also, during the ascent portion of the lift-off trajectory, the conventional control flaps would be functional. There is also an enhanced control capability with a propeller-powered Tail Sitter as the airflow will flow over the control surfaces, increasing their effectiveness.

But like any type of VTOL, there have long been inherent control problems with this concept through the years, and in most cases, other types of control augmentation had to be added.

There have been a number of Tail Sitter concepts through the years, but none ever met with any great success and none ever reached operational status.

During World War II, the Germans had considerable interest in the concept. Several test programs were initiated with the Fock-Wulf Triebfluegel being the most promising. The concept employed a wingless aircraft using a large triple-blade propeller in the mid-fuselage position. A ramjet was located on the tip of each blade, and had the revolutionary concept reached production, it could have had an effect on the end of the war. It never flew. In the U.S., there were also some gangly efforts for the straight-up concept, with Hiller Aviation designing a flying platform research vehicle. It certainly didn't look like an aircraft of any type, but it was another attempt at proving the concept.

Looking more like the Apollo Lunar Module, the Rolls-Royce Flying Bedstead research craft derived its power from a pair of Nene turbojet engines with air nozzles mounted on outriggers for vehicle control. There was about four tons of thrust available from the engines, which was about a half-ton more than the weight of the vehicle. Surprisingly, the engines on this testbed were mounted in the horizontal position, with the thrust turned through 90 degrees using unique tailpipes. The pilot location was the only place it could be, i.e. right on top of the contraption.

A Russian effort (the Turbolot) with the concept looked more reasonable with a turbojet engine mounted vertically, directly in the middle of the machine. There were deflectors in the thrust stream that provided control for the vehicle, along with compressed air nozzles that were mounted at the end of four outriggers. Actually, little is known about the so-called Turbolot, but it did indicate that the Soviets too were interested in the concept.

And when you get right down to it, the 1960s Bomarc air defense missile, which was launched vertically sitting on its base, could be classified as a Tail Sitter. Carrying either a liquid or solid-propellant booster integrally in its body, the Bomarc (which was initially designated the F-99) followed a vertical trajectory to altitude before rolling over to a horizontal flight path and lighting its ramjet sustainer motor. Even though the concept never met with much success, the concept is still

One of a number of "Flying Bedsteads" that helped set the scene for Tail Sitter VTOLs. (Holder Collection)

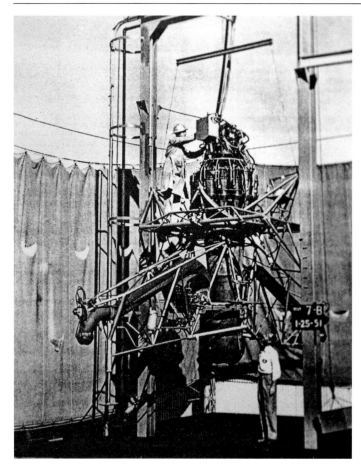

The Allison-powered VTOL test rig was used by Ryan in 1951. (Ryan Photo)

not dead, with several Tail-Sitter concepts being considered for the 21st Century. The only difference with these modern Tail Sitters is the fact that they will probably be unmanned, i.e. UAVs (Unmanned Aerial Vehicles), with the purpose of performing surveillance missions. Systems under consideration go by the project names of Guardian (from Bombardier) and the Vigilante (by SAIC).

The Guardian is a six-foot-high Tail Sitter UAV craft that weighs only 300 pounds empty and is able to carry a 200-pound payload and remain on station for up to six hours. Sporting a pair of triple-bladed props, the Guardian is powered by a gas turbine engine.

The Vigilante is actually a derivation of a two-seat helicopter. The 600-pound empty weight of the system is amazing when you consider that it could carry up to a 1,200 pound payload and fly to up to 20 hours. A manned version of the Vigilante has also been considered.

Looking like a flying saucer, the Sikorsky Cypher Tail Sitter UAV also received considerable attention during the early 1990s. First flown in 1992, the Cypher was powered by a 52 horsepower Alvis rotary piston engine. The tiny craft weighed only 175 pounds with a maximum payload capability of 45

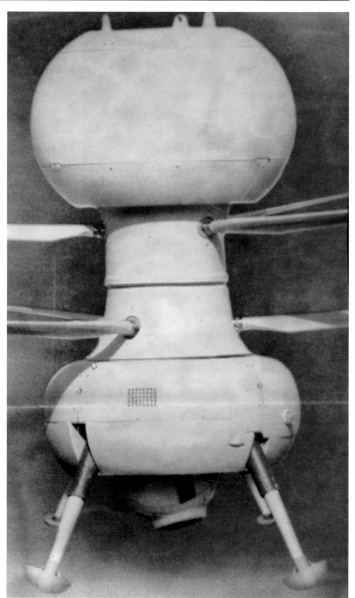

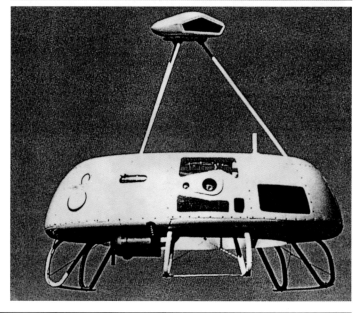

Above, Right: The Guardian UAV Tail-Sitter VTOL. (Guardian Photo)
Right: Sikorsky's Cypher, a multi-mission reconnaissance UAV. (Sikorsky Photo)

pounds. With a cruising speed of 80 knots, the Cypher had a ceiling of 5,000 feet and a three-hour flight duration.

The U.S. Navy has again been urging the embracing of unmanned vertical take-off systems for strike missions. A late 1990s Navy study indicated the advantages of such a craft which would be "Freed from the requirement for catapults and arrested landings." The Navy would benefit from such a system since the system could be employed with non-carrier battle groups.

The requirements for such a Tail Sitter system would be to reduce the level of technology required for the system since there would be no need for a pilot life support system.

Such a system is the unmanned Bombadier CL-327 Guardian, looking like something from outer space with a twin-bulbous shape and twin rotating propellers. With a surveillance mission, the strange little craft can carry a 200-pound payload and stay on station 120 miles from base with a speed capability of 70 knots.

In 1999, the Tail-Sitter concept evolved again in the form of the so-called "Atmospheric Test Vehicle (ASTV)," a vehicle which would test a single-stage to orbit launcher concept. The conical 63 foot-tall vehicle has a S-58 helicopter rotor on top.

The Tail Sitter concept will be very much alive in the next century, manned or unmanned. Following are some of the hugely-interesting Tail Sitters that met with some success, mostly during the 1950s.

LOCKHEED XFV-1 TAIL-SITTER VTOL FIGHTER (U.S.A.)

System: Built-from-scratch prototype
Manufacturer: Lockheed
Sponsor: US Navy
Operating Period: 1950-to-1955
Mission: To develop a Tail-Sitter VTOL system for shipboard use to provide air defense for the fleet.
The Story:
As early as the 1940s, the Navy had the possibilities of a small VTOL Tail Sitter aircraft. Navy officials had originally thought about such aerial vehicles for target drones. A study by the Bureau of Aeronautics determined in 1947 that such a craft could have fighter applications, and the idea took on increased priority. The study also indicated that a turboprop propulsion system would be the appropriate engine.

Early in 1950, the Navy decided to go forward with the concept and put out a call for proposals to the industry. Five manufacturers responded, with the final contracts being awarded to Convair and Lockheed, the XFV-1 being the program designation given to the latter. The contract also specified that both aircraft would use the same Allison T40 powerplant, which was basically a pair of Allison T38 engines tied into a single gear box to drive Curtiss counter-rotating propellers.

Primary consideration for the pair of VTOLs was light weight and adequate power. The counter-rotating props had the advantage of eliminating the high torque effect of a single propeller.

A service structure is wrapped around the XFV-1 Tail-Sitter. (Lockheed Photo)

The XFV-1 Lockheed design featured four identical tail surfaces and twin forward wings. (Lockheed Photo)

Design goals for both programs looked for a take-off speed of from 60-to-80 miles per hour with the capability of rising vertically to 10,000 feet. The Navy also desired a range of a thousand miles with a full fuel load. Finally, a forward speed of 500 miles per hour, the same as the T40-powered A2D Skyshark, was deemed a requirement.

This section therefore looks at one-half of that Tail Sitter contract addressing the Lockheed candidate. In retrospect, it must be noted that there was a close similarity between this system and the projected Nazi Focke-Wulf VTO fighter of 1944. Lockheed would admit that this was one of the riskiest programs it had ever undertaken.

Even though the two programs were confined to some strict rules in that both were required to use the same powertrain arrangement and same 16-foot diameter propellers, there were considerable differences in the approach that each company took with their particular program.

The YT-40 engine provided over 5,000 pounds of thrust in the lift-off mode. There were reports that the lift-off thrust might later be augmented with JATO boost rockets to increase stability during the lift-off phase.

The props of the Lockheed design, which were contained in a huge elongated spinner, were located amazingly close to the small bubble fuselage, and accounted for about one-fifth of the total length of the fuselage. Their location must have provided for a loud and shaky ride for the single pilot.

There was considerable concern by the Navy on the safety of the pilots for both the designs. As such, a parachute

was developed to float the pilot away from the plane as low as two hundred feet in an emergency situation. Equipped with a standard ejection seat, the parachute had to be able to operate at low altitudes and low speeds.

The XFV-1 fuselage was 37 feet in length, pointed on both ends, with the rear mounting a unique four-tail surface configuration. The initial purpose of those surfaces was to serve as a mounting platform before lift-off. Small landing gears were mounted on the wings' outer tips. The overall tail height was 24 feet, with a wingspan of 26 feet, and the Pogo's empty weight being only 14,000 pounds. Early in the program, the Lockheed craft was equipped with a conventional landing gear before it was moved to the vertical position for take-off.

For pure vertical take-offs, the large control surfaces on the wing caught the direct flow of the twin props, thus aiding control while the plane was in the ascent stage. Pitch control during ascent was provided by electric pitch systems in the propellers. Still, though, flight testing would show that control, especially in the hover stage, would be one of the real problems faced by this, and also, the Convair system.

Test pilots recalled the toughest maneuver with the XFV-1 was the tricky landing maneuver. There had to be near preciseness on the controls. During the landing maneuver, the seat was maneuvered to coincide with the new altitude.

There were no provisions made for armament on the system, but with its configuration, it would have been difficult to mount guns with its large propellers and short wings. It is possible that rockets, similar to the technique used by the German Natter Tail Sitter fighter of World War II, might have been more realistic.

The flight testing of the XFV-1 commenced in 1954, but amazingly, even though there was considerable money invested in the program, there were still no vertical lift-offs or landings ever attempted. A total of 27 conventional flights were made.

Since conventional take-offs and landings were made by the craft, it was necessary to fit the plane with a somewhat conventional style landing gear located directly under the cockpit. The first conventional flight using that landing gear was accomplished in March 1954.

In level flight, the plane had the look of a homemade machine, appearing ungainly in flight at best. Maybe it was that appearance, which was about as far away from the look of a fighter as you could get, that might have influenced its eventual cancellation.

Actually, though, it came down to the control problems experienced during hover flight which presented real problems for the test pilots. The problem came from the situation of not being able to determine sink, climb, and rotation in a normal mode of flight.

The program was officially canceled in 1955.

A rear view of the XFV-1 showing the rear wings. (Lockheed Photo)

CONVAIR XFY-1 POGO TAIL SITTER VTOL FIGHTER (U.S.A.)

System: Built-from-scratch prototype
Manufacturer: Convair Division, General Dynamics Corporation
Sponsor: US Navy
Operating Period: 1950-to-1956
Mission: To prove the concept of a small Tail Sitter fighter concept to be launched from US Navy ships.
The Story:

The Convair XFY-1 fighter was the second of a pair of prototype programs to develop a mini Tail Sitter fighter which could be deployed on non-aircraft carrier-type ships, the other program being the Lockheed XFV-1 Pogo. It was truly a prototype fly-before-you-buy situation, but unfortunately, there would be no winner in this competition, as production was not in the future for either of the very similar fighter designs.

Testing for the fledgling Convair design was begun with the completed prototypes in 1954, with testing in a special tethered-flight rig amazingly installed in an airship hangar at Moffett Field, California. Some 300 flights were made.

Convair came up with the idea of tethered flight when it learned it could make use of the Moffitt Field hangar, which had been built in 1933 to shelter the *USS Macon* airship. Its immense size enabled the company to install a 184-foot rig where a system of cables were strung through pulleys and hooks to an electrically-controlled system to control the plane during its vertical testing. The device also served to restrict the lateral travel of the strange little craft.

Following the extensive tether testing, the XFV-1 would perform its first outdoor flight in August 1954, and would rise to a height of only 20 feet. Later, other flights would see an altitude of 150 feet.

The first full VTOL maneuver took place in November of the same year. The operation consisted of taking off from the pure vertical attitude, then tilting forward to accomplish the touchy transition to horizontal flight, and then accomplishing

The squatiness of the XFY-1 Tail-Sitter is evident from this angle. (Convair Photo)

the exact opposite operations to end with a tail-down vertical landing. The first transition accomplishment was met with much adulation, since the effort was the first such transition from a non-helicopter-type VTOL vehicle.

But the excitement would be short-lived, as the program would experience similar problems as its Lockheed brother. The program experienced considerable propeller and engine problems in the tests that would follow.

But suddenly, the Navy was starting to look unfavorably on the Tail Sitter concept as a whole. One of the main problems it considered with the concept was a maintenance problem because of its vertical attitude making the higher points of the plane difficult to service.

Then, there was also the situation of pilot position. Granted, there was an adjustable seat that rotated the pilot 45 degrees forward when the plane was in the vertical attitude, but there was still the situation of the pilot being on his back with his feet sticking up in the air. Normally, the canopy was open during the ascent phase of the trajectory.

It came down to the situation in landing that the pilot was forced to peer over his shoulder at the ground. Still, the test pilots never had that first misfire in all the landings that were made, certainly a testament to their obvious skills.

Probably almost as difficult as flying the Pogo was getting into the machine. Sitting about two-thirds of the way up the fuselage, while in the vertical attitude, a special ladder was required for the pilot to assume his position. Interestingly, there were 13 steps in the ladder to make the trip.

Even with the orientation situation, pilots in the program were impressed with the ease of operation of the stubby little

The Convair XFY-1 Tail-Sitter in flight. (Convair Photo)

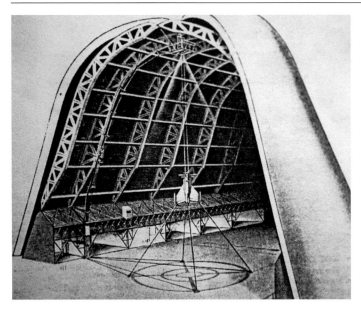

This drawing shows the special tethered flight rig used for testing the Convair XFY-1. Note the elaborate cable system. (Convair Drawing)

Note the difference in size of the pairs of rear wings. (Convair Photo)

fighter. The amazing aspect of the Convair product was the fact that there were no additional devices for stabilization during the ascent and descent.

There was considerable dimensional similarity between the Convair and Lockheed Tail Sitters. The Convair model showed a maximum height of just over 32 feet tall—that's standing straight up of course. The wing span was 27 feet, eight inches, and the wings carried a slight positive dihedral angle. The total wing area was 355 square feet, with fin and elevon areas of 176 and 36 square feet, respectively.

It carried a minimal gross weight of only 14,250 pounds, which was considerably less than the Lockheed design and aided materially in its performance.

Positioned at 45 degrees to the wings were the two tail surfaces, one located directly behind the cockpit with the other directly below on the lower fuselage. Auxiliary fuel tanks were carried on the tips of both wings and were much thinner than those used on the XFV-1. Also recall that the Lockheed Tail Sitter had four tail surfaces with the wings in a more conventional location.

The wings and tails also served an additional purpose of mounting small caster wheels on their tips which all together served as the landing gear. An interesting innovation on the lower tail was the fact that it could actually be jettisoned should the occasion arise that the plane had to land in a conventional manner.

There was nothing unique about the control surfaces, as the trailing edge of the wings and tails provided the necessary control during the vertical take-off phase, conventional horizontal flight, and also during the landing phase. Their effect was actually enhanced during the hovering phase because the surfaces were a direct recipient of propeller airstream.

The fuselage had an interesting shape to it, being extremely bulbous in the cockpit area and tapering down abruptly at the rear.

To keep everything pretty equal between the two contractors, it was decided that both would utilize the same Allison YT40-A-14 engine, with a horsepower rating of 5,850, driving a pair of Curtiss-Wright 16-foot-diamter counter-rotating propellers.

The twin-turbine YT-40 had a high-speed, lightweight reduction gear system with a capacity of more than seven thousand shaft horsepower. A speed reduction ratio of 14-1 was necessary in order to reduce the turbine rotational speed to a usable propeller speed.

The nose, which contained both propellers, was longer and more streamlined than the Lockheed design. Had either design reached an operational status, the Navy had plans on outfitting the winner with the more-powerful Allison T54 engine. The intakes for the engine were situated in bulges in the wing roots. The top speed of the model was quoted at about 500 miles per hour. Although it didn't receive the publicity of the actual flights, there was also an interesting tee-pee-shaped hangar built specifically for the Pogo. The "hangar" consisted of two portions which were moved together by tractors. They were then joined together like a clamshell around the plane. Also within the hangar were service platforms located on both sides of the plane which allowed easy access for maintenance operations.

With some promise, there were hopes that the Pogo program would continue, but the program would be canceled along with its running mate Lockheed entry.

It wasn't all for naught, though, as there was much learned about the functioning of the control surfaces in the slipstream and the characteristics of the vertical-to-horizontal phases of VTOL flight.

The main reason for the XFV-1 failing to progress toward operational service was a worry about sufficient power to guarantee safe VTOL operations. It wouldn't be long before the whole VTOL Tail Sitter fighter concept would be abandoned in favor of more conventional designs.

RYAN X-13 VERTIJET TAIL-SITTER VTOL FIGHTER (U.S.A.)

System: Built-from-scratch prototype
Manufacturer: Ryan Aeronautical Company
Sponsor: US Navy, USAF
Operating Period: 1947-to-1957
Mission: To develop a Tail Sitter VTOL which would utilize a typical vertical lift-off, transition to horizontal flight, and then descend vertically to be engaged by a line which engaged a hook on the nose of the plane.
The Story:
The story of the Ryan X-13 actually started following World War II, when the Navy had a huge interest in fighter aircraft that could be launched vertically from non-aircraft carrier type ships. The designated mission of these craft was to be close-in air defense.

Experience had shown that, during the war, the massive aircraft carriers were such a prime target that being able to scatter out the fighter aircraft could provide a great tactical advantage. Even submarines were considered as possible launching ships should a small-enough Tail Sitter be developed.

"Compact" would have to be the prime descriptor of the Vertijet. (Ryan Photo)

Unlike the Convair and Lockheed double counter-rotating propeller Tail Sitter vehicles, Ryan concept studies would use a pure-jet propulsion system, and the company brought much experience to the program, having just developed the Ryan FR-1 Fireball fighter for the U.S. Navy.

Knowing the Navy interest in small VTOL Tail Sitters, Ryan had earlier proposed a system using the hard-to-control pure jet system applied to a Tail Sitter VTOL.

Before there could be a serious proposal made, it would be necessary to address the control problem, so Ryan considered and tested a pair of jet reaction systems. One of the systems used scoops inside the engine tail pipe to pass a

The X-13 slowly rises in one of its first flight tests. (Ryan Photo)

Test rig for the X-13. (Ryan Photo)

Ryan test pilot Pete Girard. (Ryan Photo)

Launch Platform for the Vertijet. (Air Force Museum Photo)

small amount of exhaust gas for yaw control during the critical hover stage. For roll control, the engine nozzles were actually rotated opposite of each other.

Technique number two separated the main tail pipe just aft of the engine into a pair of parallel tail pipes. Each pipe used an "eyelid" to modulate the thrust. Pitch control was achieved by rotating elbows in each tailpipe. Roll control was a complicated technique with the rotating elbows rotated differentially to achieve that effect.

In 1947, Ryan presented a formal proposal to the Navy, and the Navy responded with a contract for the princely amount of $50,000. Following a test program proving the concept, the Navy contract was continued in September 1948 for the amount of $202,830.

Detail of X-13 vertical tail. Plane is on display at Air Force Museum. (Bill Holder Photo)

Horizontal tab located on each wingtip of the X-13. (Bill Holder Photo)

The curved top of the X-13 wing. (Bill Holder Photo)

The X-13 had a unique shape no matter from what angle it was viewed. (Air Force Museum Photo)

Before the actual flight test vehicle could be built, though, there were a series of ground tests performed using a test rig fitted with a delta wing, refined jet reaction system, and a cockpit which had been fashioned from a scrap B-47 fuel tank. The devise was powered by an Allison J-33 turbojet. The pilot controlling the awkward craft seated in a tilted plywood seat.

That, though, was the final configuration of the rig, having started out as not much more than a simple tubular structure with the J-33 engine mounted in the center. It looked about as far from an aircraft as could be imagined. This initial rig first lifted off the ground in 1950, and a year later, made its first remotely-controlled hovering flight.

This series of tests proved that the Tail Sitter concept being considered appeared to be feasible, but there were changes required for the actual flight test vehicles.

The controllability of the craft was found to be extremely good, but it was determined that with a more-powerful engine, the response could be greatly increased. The fact that the craft was sitting considerably above ground level meant there was very little of the expected ground effects or high gas recirculation experienced. A positive result of the testing

A demonstration flight for the military community in the Washington DC area. (Air Force Museum Photo)

X-13 Vertijet shown during the landing maneuver attaching to the retrieval wire. (Air Force Museum Photo)

X-13 in a test rig. (Ryan Photo)

Check the way the pilot is leaned forward in the X-13 as he awaits lift-off. (Ryan Photo)

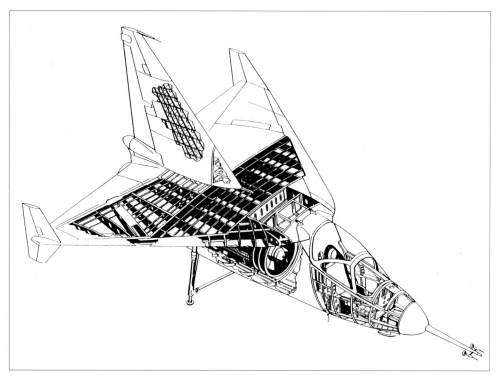

Cutaway drawing of the X-13. (Ryan Photo)

The X-13 pilot found himself in a strange attitude during the lift-off phase. (Ryan Photo)

showed that, with experience, the pilot skills could be greatly increased.

The final vehicle configuration that evolved through this research was typical of the Tail Sitter shaping of the period, being best described as "Stubby." The X-13 prototypes were very small, with only a 21-foot wing span with a fuselage length of 24 feet. The top of the tail stood at 15 feet. The plane's gross loaded weight was 7,500 pounds with a 1.3-1 thrust-to-weight ratio.

A sharp 60-degree wing sweep was selected for the craft since Ryan engineers felt it would possess good stall characteristics. The large vertical fin, fitted with end plates, provided for directional stability, while the elevons accomplished pitch and lateral control.

The powerplant for the final flight vehicle was a single Rolls-Royce Avon turbojet powerplant capable of ten thousand pounds of thrust.

The go-ahead for the building of the first prototype occurred in 1953 with the first horizontal flight taking place in December 1955. In May 1956, the first Tail Sitter straight-up lift-off would occur with the first prototype, while the second craft would make the same flight six months later. The existence of two prototypes added a significant amount of security to the program, since it wouldn't have been completely shut down in case of the loss of one of the aircraft.

In April 1957, the first complete flight sequence was demonstrated with a vertical take-off, transition to horizontal flight, and a return to a vertical landing being captured by its unique launcher. The accomplishment was made by the second prototype.

The pair of prototypes were initially fitted with a temporary landing gear for the early pure-horizontal landings, but later in the test program it would be replaced with a rear-mounted framework arrangement.

Its that launcher-recovery device that deserves mention, as the X-13 was the only Tail Sitter to utilize such a technique. The launcher, in fact, also served as a landing platform, as a single hook mounted under the X-13 forward fuselage held the X-13 in position during launch and then re-engaged the craft during the landing maneuver.

The device was actually a horizontal cable extending between upright supports of the launcher. If that sounds like it might have required some precise landing maneuvering, that is a very good assumption.

Watching the complete landing transition of the X-13 was an amazing experience, seeming to defy all the laws of physics and aerodynamics. First, the vehicle would tilt slowly from the horizontal to vertical flight, then momentarily stand almost motionless in the upright position in mid-air. From that position, it would then slowly move in to hook on the stretched cable with its nose hook.

The device had some positives from the viewpoints of both services. For the Air Force, it was visualized that, lacking runways, the launcher could be moved into position almost anywhere, where it quickly could be prepared for vehicle launch. The Navy, of course, saw it differently, where the apparatus could be mounted on the side of carriers, or on transport ships without decks, but with enough space to store the required launching apparatus.

In June 1957, the X-13 was given a dramatic opportunity to demonstrate its capabilities where it hovered across the Potomac River to the Pentagon. It was all to no avail, though, for the Vertijet, as the Air Force decided, for some reason, that it would not continue the development.

Undoubtedly, the earlier cancellation of the Lockheed and Convair Tail Sitters had to play in the demise of the X-13.

SNECMA C.450 COLEOPTERE TAIL-SITTER VTOL (FRANCE)

System: Built-from-scratch prototype
Manufacturer: SNECMA
Sponsor: Company program
Operating Period: 1952-to-1959
Mission: Provide an initial investigation of the Tail-Sitter VTOL concept.

The Story:

The Tail-Sitter concept is one of the most radical of the straight-up concepts, requiring a push of the state-of-the-art technology of the 1950s time period. The technique attempted by the French SNECMA organization to accomplish the technique evolved into an extremely interesting vehicle configuration.

Research on the possibilities of this unique VTOL concept actually started in the late 1940s, with Dr. Ing. Hermann Oestrich doing considerable investigations before assuming his position as Engineering Director of SNECMA. The company then initiated a number of prototype programs, the first being the so-called C.400 Atar Volant, of which several versions were built. The P-1 version was a pilotless remotely-controlled vehicle with an Atas 101 turbojet equipped with a jet-deflection nozzle for control.

Resting on its transporter, the pilot checks out the cockpit before flight. (SNECMA Photo)

An early testbed for the Coleoptere. (SNECMA Photo)

The P-1 looked much like a hay silo mounted on a mobile tubular four-wheel undercarriage. An annular fuel tank was fitted around the center of the nacelle. Control was supplied by a gyroscopic system.

The P-2 system, this time carrying a pilot, was similar to the P-1. An extensive flight test program was conducted in 1957 and 1958 with both tethered and free-flight tests. The technology continued to increase with the P-3 version, which incorporated a more powerful engine. In a wind tunnel simulation, the system was mounted atop a railroad truck, and towed at speeds up to 50 miles per hour to test the effect of airflow on the engine jetstream during a rapid vertical descent. Those and other research efforts would all lead to the C.450 Coleoptere project, whose outward appearance certainly got curious attention. The "plane," if you can call it a plane, didn't have a conventional wing, instead an annular 10.5 foot diameter device that completely encircled the body.

Four small fins were attached to the "wing" and dropped below the bottom of the shroud. The vehicle was mounted on four small castor wheels, with rubber tires, which provided the landing platform.

Jet power was initially in place in the form of a 6,400 pound thrust Atar D jet engine, but later came the C.450 8,155 pound thrust Atar 101.E turbojet. The C.450 actually used the Vectored-Thrust concept in controlling the vehicle during its lift phase with movable vanes in the engine thrust stream.

As with all Tail Sitter VTOL vehicles, directional control was a huge concern. That control at take-off and landing was accomplished by deflection of the main exhaust, while during horizontal flight, control was maintained by four swiveling fins at the rear end of the wing. During transitions, small fins on the outer fuselage aided in the process.

The 23-foot-long fuselage appeared to be emerging from a tunnel with about half its length above the top of the annular wing. Two small retractable strakes in the nose were in place to accomplish a pitch-up moment during the transition phase to horizontal flight. The complete airframe was built by Nord Aviation and fabricated of a light alloy construction.

The estimated horizontal speed was about 500 miles per hour, with a rate-of-climb of almost 26,000 feet per minute and a service ceiling of about 10,500 feet.

The 6,600 pound C.450 made its first flight test in vertical flight in May 1959. Then, on July 25th, during the vertical-to-horizontal transition phase, the pilot lost control of the craft at 250 feet altitude. Fortunately, there was a successful pilot ejection.

The annular wing concept could well be the ultimate design for the Tail Sitter concept and could well have advantages over the stubby Convair and Lockheed designs. The designers also considered future versions of the concept where the center cylinder could serve as a part of a ramjet propulsion system.

11
Zero-Length Launch Technique Systems

It might be questioned why the Zero-Length-Launch technique should be a part of a book on VTOL systems, but the technique actually produces the same effect as an aircraft that achieves the straight-up effect completely using its own propulsion system.

Think about the final result of having a VTOL system. The plane leaves the ground and achieves altitude from point A with no requirement for a runway. Well, the same could be stated for the Zero-Length-Launch technique, which achieves the same effect with the aid of some significant ground support equipment.

It could be stated that the concept seems more applicable to use with air-breathing missiles, but it has also been tried with great success with fighter aircraft.

The technique involves mounting either the aircraft or missile on a launch platform. The launcher, though, does not launch the vehicle at a vertical position. Instead, the launch occurs at a shallow angle, boosted from the launcher by one or two rocket boosters. When the vehicle is being launched, its engine(s) are running at near full throttle to aid in the launch. Then, once the rocket boosters have depleted their propellant, they are jettisoned.

During the Second World War, Germany developed the Natter, a manned, vertical launch interceptor aircraft. After the war, the United States developed numerous unpiloted interceptors and long range bombers. The U.S. Air Force expanded this capability by conducting tests using F-84 and F-100 aircraft and proving that the technology developed for launching unpiloted missiles could be transitioned to piloted fighters. In retrospect, it must be blamed on international politics of the cold war that this system only became operational with air-breathing missiles. A number of systems achieved operational status using the technique, including the USAF Snark, Matador, and Mace, and the Navy Regulus.

ZERO-LENGTH-LAUNCH MISSILE SYSTEMS (U.S.A.)
Northrop Snark SM-62
Even though the Snark certainly had a different appearance than current intercontinental missiles, looking more like an aircraft than a missile, it was definitely a missile.

The Snark was a missile with intercontinental range, but it was accomplished basically with an air-breathing propulsion system which was used after a solid-propellant-rocket boost stage. The nuclear-capable Snark was mounted on a

Head-on shot of a Snark on its launcher awaiting launch. (Air Force Museum Photo)

The Snark resembled a manned aircraft in many ways. (USAF Photo)

A Snark being towed behind its tow vehicle. (USAF Photo)

Matador missile on zero-length launcher. (Bill Holder Photo)

wheeled mobile launcher that appeared small compared to the size of the 69 foot long, 42-foot wing-span system.

Excluding boosters, the Snark had a launch weight of 51,000 pounds, including 22,000 pounds of onboard propellant and 5,000 pounds of fuel carried in wing-mounted tanks. The system had a reported range of 6,300 miles with a maximum altitude of 60,000 feet.

The Snark's self-contained Pratt & Whitney J57 turbojet engine provided 10,500 pounds of thrust, which carried the missile to the target at a high subsonic speed (Mach .94), making it a somewhat-lucrative target for modern fighter aircraft.

An initial version of the system, designated the N-25, was powered by a smaller J-33 turbojet engine which had a maximum range of 1,550 miles and a Mach .85 cruising speed. The final configuration was known as the Super Snark and carried the N-69 designation.

Each Allegany Ballistics Lab solid booster on the final system configuration provided 130,000 pounds of thrust to propel the system off its launcher. In today's world, the Snark would probably be classified as a cruise missile.

The Snark's trajectory was interesting in that it was fitted with a jettisonable warhead, which was blown clear of the airframe at the end of the cruise phase. The warhead then proceeded to the target in a ballistic free-fall. It also carried a non-jammable inertial guidance system capable of operating in all weather conditions and at night.

From the beginning of its design phase, the Snark system was designed to be air-transportable and was compatible with both the C-124 and C-133 transport aircraft.

An extensive flight test program was carried out from Cape Canaveral—more than 80 total flights—with many of them being fired to the full-length range. The program was, however, plagued with numerous problems.

The system, nevertheless, assumed operational status in mid-1959 with the 702nd Strategic Missile Wing at Presque

Isle AFB, Maine, a Strategic Air Command Base. Its service time, though, was minimal, being phased out in 1961. Besides its slow speed, other reasons given for its demise were the vulnerability of its above-ground launching system, low reliability, and inability to penetrate.

Martin B-61 Matador

The requirement for the Zero-Launch-Length Matador was established in August 1945. The specifics indicated a need for a supersonic version, but that aspect would later be

Solid booster motor on Matador missile. (Bill Holder Photo)

Details of the forward fuselage of the Matador missile on display at the Air Force Museum. (Bill Holder Photo)

This launch was part of the early tests of the Matador at the Missile Test Center in Cocoa, Florida. (Martin Photo)

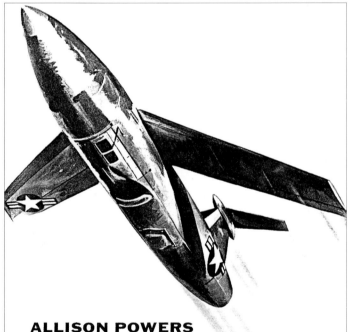

ALLISON POWERS THIS PILOTLESS JET BOMBER

Already on NATO assignment with the 12th Air Force in Europe, the Allison J33 Turbo-Jet-powered Martin B-61 Matador is an outstanding example of precision engineering.

Since its whole life involves only one "one-way trip" — both engine and missile were skillfully designed to deliver short-haul reliability at rock-bottom expenditure of man power, materials and money.

As America's first pilotless jet bomber, the Matador is a prime example of how Air Force and Industry work together to serve nation's security, economy and well-being.

Matador delivers more defense per dollar, uses fewer materials — uses electro-mechanical control in place of the pilot and crew.

GENERAL MOTORS

Allison

Division of General Motors, Indianapolis, Indiana

World's most experienced designer and builder of aircraft turbine engines — J35 and J71 Axial, J33 Centrifugal Turbo-Jet Engines, T38, T40 and T56 Turbo-Prop Engines

A period Allison advertisement of its J33 Turbojet that powered the Matador. (Allison Advertisement)

One of the Matador's strong points was its mobility on this mobile launcher. (Martin Photo)

Desert launch of a Matador. (USAF Photo)

Matador at the ready. (USAF Photo)

The solid rocket booster for the Mace missile. (Bill Holder Photo)

dropped, and the Matador missile which would evolve would continue the same subsonic performance as the Snark. But even so, the program went forward, with over one thousand of the systems built.

It's interesting that the Matador was initially given the B-61 designation, the same "B" designator as was used for manned bombers. But of course, in this case, the system was unmanned. The Air Force would later designate the system, TM-76A.

The Matador would be operational during the 1950s and equipped units in the United States, Taiwan, and Germany. The first operational USAF unit was the 1st Pilotless Bomber Squadron in 1951. The unit went to Germany in 1954, and reached operational status the following year. In Korea, the 58th Tactical Missile Group became combat ready with the C version in 1959, and remained in place in 1962. In 1961, the 498th Tactical Missile Group was deployed in hardened sites on Okinawa.

There were two versions of the nuclear-capable Matador, the first being the TM-61A, which had a MSQ radar guidance system. There was a network of ground radar stations which accomplished line-of-sight control of the Matador in flight. The TM-61C was an improved version of the A version with a new guidance system which transmitted a long-range hyperbolic grid from two transmitters, one of which controlled the missile in azimuth, with the other controlling range.

The Matador Zero-Launch-Launcher consisted of a specially-constructed semi-trailer which was developed by Martin and the Air Force. The 35,000 pound device could operate in a one-hundred square foot area and was built by the Fruhauf Company.

The Matador design featured high-mounted swept wings, a cylindrical all-metal fuselage, and a T-tail configuration with a movable rudder. The overall length of the system was almost 40 feet, with a maximum fuselage diameter of four-feet, six-inches, and a 28-foot, seven-inch wing span. Its cruising speed was an impressive Mach 0.9 with a range of over 500 miles.

Its solid booster firing, this Matador comes off its mobile launcher. (USAF Photo)

Mace missile on launcher. (Bill Holder)

The Mace B was an improvement over the initial version shown here being launched out of a bomb-proof shelter. (Martin Photo)

Martin Mace

The Mace system actually started out as the B version of the Matador, but then acquired a different designation and became a program on its own.

Renamed the TM-76A, the Mace could be recognized from the Matador with its longer fuselage and more-rounded nose shape. The Mace was sleeker than the Matador, with shorter-span wings that were designed to fold instead of being to removed for transit.

Power was more significant, with its 5,200 pounds aided for both versions by an Allison J33 turbojet with a Thiokol solid-propellant booster with 97,000 pounds of thrust.

The Mace was carried on an impressive mobile launcher built by Goodyear Aircraft and pulled by a tractor support unit. All the vehicles had large "pillow-tires" which enabled significant off-road capability.

Of the two versions of the Mace, the TM-76A carried a Goodyear ATRAN guidance system which compared a film strip to the ground it was passing over. Should the missile deviate from the programmed trajectory, its trajectory was immediately corrected. This version could also be programmed to make changes in its altitude, which allowed it to attack at a very low altitude.

Mace production at a Martin facility. (Martin Photo)

A Mace is partially hidden in a tactical environment. (USAF Photo)

A number of Mace missiles are lined up on their launchers. (USAF Photo)

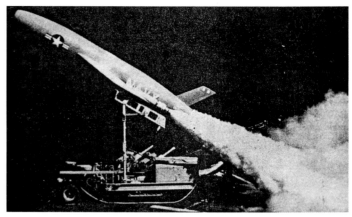

The portable launcher of the Navy Regulus is shown being launched off the deck of the carrier *Hancock*. (Chance Vought Photo)

Chance Vought Regulus

The result of a late-1940s requirement for a simple subsonic missile system, the U.S. Navy Regulus first took to the air in 1951. It remained operational in a shipboard environment for a number of years with 514 being produced. The initial Regulus (there were two versions) was deployed on both cruisers and submarines.

Like the Matador, the Regulus suffered from a number of technical problems during its development phase. It came down to the fact that, like the Air Force, the Navy probably preferred a manned aircraft as opposed to this subsonic system.

Different from the Matador/Mace family, the Regulus featured a mid-fuselage-mounted swept wing, a single rear vertical fin, a cylindrical fuselage constructed of light alloy/balsa sandwich, a command guidance system, and a nuclear warhead. Initial boost was provided by a pair of 33,000 pound-thrust solid boosters.

Dimensions included a 33-foot body length and a 21-foot wing span. The Regulus I had a launch weight of 14,500 pounds, with a range of 575 miles and a cruising speed of Mach 0.87. The launcher was a simple steel frame structure which mounted the missile at a low-launch angle.

A second version of the missile system, the Regulus 2, provided significant performance increases over the original. The second version of the Regulus was unique in that it could make a conventional take-off with a conventional landing gear, or could be launched from the ramp launcher with rocket assist. The program was canceled in December 1958.

Manned Zero-Length-Launch Aircraft Systems
BACHEM Ba-349 NATTER ZERO-LENGTH-LAUNCH FIGHTER (GERMANY)

Aviators and designers had thoughts of using rockets for propulsion almost from the beginning of aviation. The first known attempt to mate solid-propellant rockets to a glider was made in Germany in 1928. In 1929, an overloaded Junkers W33 seaplane took off from Elbe River near Dessau, Germany, using six black-powder rockets.

Germany continued such research right up to the closing days of the Second World War. Quite simply, the Natter Ba-349 could be called a weapon development of desperation. The war was not going well for the Axis, and the cry went out from Adolph Hitler to develop a family of "Secret Weapons." Without doubt, had the call gone out earlier, say in 1942, the cost of winning the war for the Allies could have been greatly increased.

It was one of a number of weapons that appeared during the war's closing days, including the likes of the jet-powered Me-262 and the rocket-powered Me-163 Comet. There were also many others on the board that never made it to operational status.

The pressing problem facing the Third Reich was the unending stream of Allied bombers obliterating the German landscape, and a faltering Luftwaffe that was running out of planes. What was needed was something cheap, fast, and capable of bringing down the bombers.

The order was a tall one, and time was wasting, when the task was given to Herr Knemayer to develop the Natter. The design was about as simple as you could get. Get a small rocket-powered aircraft into the path of the bombers with rocket power, and then let the pilot take over and fire a volley of unguided rockets into the fleet.

A Natter on display at an airshow. (Holder Collection)

The remarkable aspect of the so-called Ba-349 Natter program was the length of its design and development program, a scant six months. Needless to say, the air of urgency was a great motivator. Reportedly, the staff for the program was only 300, with about 60 being engineers.

Theoretically, the Natter can't really be classified as a VTOL, since no provision was made for landing.

Initiated in August 1944, which was just after the Allies' invasion of France, the first flight test was accomplished in late February 1945. The war was, for all practical purposes, over, but the program would continue undaunted.

That first flight test, though, met with ultimate destruction, spreading parts from the unique craft over the German countryside and killing the test pilot, Lothar Siebert. But the program went on with gusto, and within the following two months, the craft had been successfully launched, and more importantly, the pilot was successfully recovered.

With no time to wait, it was necessary for the Germans to quickly deploy the tiny craft to slow the oncoming of the inevitable. In probably the quickest program in the history of aviation, ten of the Ba-349s were deployed in an operational squadron, but their effectiveness was minimal, and there were probably more casualties suffered by pilots of the craft than they inflicted on Allied bombers.

The typical trajectory for the Ba-349 was bizarre to say the least. Launched vertically, the rocket engine could kick the craft up to 29,000 feet in less than 60 seconds. Needless to say, the pilot had to endure some pretty significant G loadings during that quick trip.

The pilot would then take over, with the first job being to jettison the nose cap and begin attacking the bomber formations with his offensive package of either 24 73mm Foehm or 33 R4M 55mm unguided rockets. The attack had to be quickly initiated, with only two or three passes before the rocket propellant was expended. It was all gone after only about seven minutes, when pilot concern turned to his own safety.

No landing here—the craft was expendable, and through a crude ejection system, the pilot jettisoned himself, hopefully to a safe landing.

The procedure involved diving and then gliding down toward a safe landing location. The pilot would then free himself from his seat belts, lean forward and pull a lever which would detach the nose. Aerodynamic forces would then actually remove the complete forward portion of the plane, exposing the pilot to the direct windstream.

The pilot would then deploy his chute, which unfurled a large parachute attached to the rear of the plane. This would throw the pilot forward and clear of the structure. Another parachute would lower the remainder of the plane to the ground. It goes without saying that there were some brave pilots that flew the Ba-349.

It should be noted that during the initial portion of the launch phase, the speeds were too slow for the wing flaps on the tail to provide effective aerodynamic forces. The answer was control vanes located in the rocket engine exhaust which

would later melt in the high temperature.

A look at the Ba-349 itself reveals a small 18 foot, nine inch fuselage length and short straight wings, with only 46 square feet of area, which were only a dozen feet in total length. The main spar for the wings was of a laminated construction.

The propellant tanks were carried over the low-mounted wings, with an armored bulkhead in place just forward of their location. Additional protection was also in place directly in front of the tiny cockpit. There was a four control surface tail structure, with the top vertical member mounting a rudder. The horizontal stabilizers both carried sizable elevons.

Like the earlier British Mosquito twin-engine fighter, the Natter was also built mostly of wood and other non-metallic materials. Power was significant, acquiring about 3,700 pounds of thrust from its Wager HWK109 rocket engine. The advanced engine was capable of being throttled down to as low as 300 pounds by the pilot. Fully loaded, the Ba-348 weighed about 4800 pounds, of which 1400 pounds was the liquid rocket propellant.

To augment the lift-off thrust, the craft was aided by four solid rocket Schmidding 109-533 boosters which burned for a mere ten seconds before being jettisoned.

Interestingly, near the end of the war, there were reportedly plans to sell the plane to the Japanese, but that never came to pass.

F-84 ZERO LENGTH LAUNCH AIRCRAFT SYSTEMS (U.S.A.)

By the early 1950s, short ramps were used routinely to launch early cruise missiles. Engineers figured that perhaps this concept would work just as well for manned aircraft. But eliminating the runway for launch only solved half of the problem... one still would be needed for landing. But perhaps not. Thus was born the ZELMAL (ZEro-length Launch and MAt Landing) program. A rocket would be used to launch a fighter aircraft, then use an inflatable rubber mat, an arresting cable, and a tailhook for the landing. The mat they came up with measured 80 X 400 feet, was 30 inches thick, and had a slick surface coated with a lubricant to assure a smooth landing.

Firebee drones also used the zero-length technique. (USAF Photo)

A Firebee is zero-length launched from a ground launcher. (Air Force Museum Photo)

ZELMAL was primarily an exploratory research program. There was no specific tactical requirement for such a capability. The objective was to show increased tactical mobility and decreased vulnerability through wider dispersion and more effective concealment of aircraft in a front line area. The Air Force awarded a contract to the Glenn L. Martin Company of Baltimore, MD, to develop the system to launch a Republic F-84G and to conduct the tests at Edwards AFB. Martin already had developed the B-61 Matador surface-to-surface missile which was ground-launched using a rocket engine. Everyone agreed that the principle and techniques needed to launch a piloted aircraft would be similar to those used for the B-61.

As originally planned, the program called for the launch of an unpiloted F-84G, approximately 10 lightweight launches of piloted aircraft with conventional landings, approximately 30 launches at greater gross weights, and approximately 30 mat landings. The launches would be performed from a B-61 launcher. The lightweight launches would use a T-50 booster producing 57,000 pounds thrust for 2.5 seconds. The heavyweight launches each would use an X-224-B1 rocket booster producing 75,000 pounds of thrust.

The pilotless F-84G launch was performed on December 15, 1953. Test data indicated that the T-50 produced less thrust than expected, resulting in lower speed at burnout than predicted. To compensate for the decreased thrust, the launch angle was lowered from the original 20 degrees down to 17

degrees, flaps were lowered to 5 to 10°, and the gross weight was lowered slightly.

Preparation for the manned launches began at Edwards AFB immediately after the pilotless launch. The first launch was on January 5, 1954, using F-84G S/N 51-887 at a gross weight of 14,270 pounds and using a trailer mounted launcher. The aircraft was shown to be completely controllable, and the pilot had no problems at any time. The acceleration was comparable to a conventional catapult launch. The only minor problem was the pilot's left hand resting on the throttle. The sudden acceleration when the booster started forced his hand to yank the throttle back to 93 percent revolutions per minute. The pilot observed that by rocket burnout, the aircraft had attained an airspeed of 175 miles per hour and an altitude of 90 feet, with no trim change needed. He noted the retarded throttle and advanced it to 100 percent to complete an otherwise uneventful launch.

Meanwhile, the second test aircraft arrived at Edwards on January 22, 1954, and was instrumented for tests. This aircraft would be launched at heavier weights of up to 23,000 pounds starting in July, using the larger X-224-B1 rocket booster producing 75,000 pounds of thrust.

The second light weight launch was performed on January 28, 1954, again using 51-887. To prevent the throttle from moving as it had on the first launch, a throttle lock was installed to prevent it from slipping. The lock could be disengaged easily using a thumb switch. Instrumentation indicated an airspeed of 174 miles per hour at rocket burnout, and a slight pitch up during the rocket burn, but well within the pilot's ability to maintain control.

The fourth through eighth launches were performed during February 1954. All were made with intentional lateral or vertical misalignments of the booster to determine the limits of aircraft stability and pilot control, assuming that operational users would not be able to align the booster perfectly every time. Two more launches were performed in March 1954.

The idea of the mat landing system intrigued many people. The mat itself could be transported by just a few trailer trucks and set up on an unprepared surface. Essentially, it was an

Zero-length-launch of an F-100 fighter from a fixed facility, the only time it was attempted. (Air Force Museum Photo)

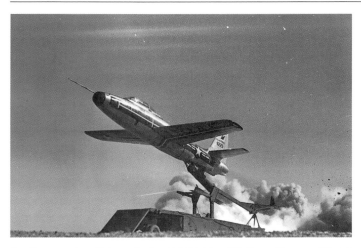

Another zero-length-launch of an F-84 fighter. (Air Force Museum Photo)

almost-instant landing field. A Navy type Mark V arresting gear was used to snatch the aircraft's tail hook.

While aircraft carrier-type arrested landings on almost any paved surface could have been performed, the Air Force did not want to strengthen their aircraft to accomplish true carrier type landings (the tailhooks on Air Force aircraft are designed for emergency use only and are not as strong as on Navy aircraft). This would have required a stronger landing gear and internal supporting structure for the gear and arresting hook, and redesigning many systems to withstand the higher longitudinal forces. By landing on the mat with the arresting gear adjusted to produce a gentler force, no structural changes, with their resulting weight increases, would be needed. In addition, the need for a landing gear could be eliminated all together. This would save 500 pounds that could then be used for additional weapons or fuel. The mat itself was made by the Goodyear Tire and Rubber Company.

The mat leaked on the first try at inflating it to the intended pressure of 4 pounds per square inch. Several of the cells were sent back to Goodyear for repair. In preparation for the first landing, several low passes over the mat were performed. These were referred to as "bounce tests," and it was decided to use water as a lubricant for the first mat landing. However, the available documentation did not specifically state if actual bounces off the mat were made during these tests. At least one snatch of a clothesline was made at 160 miles per hour to gather data on the tail hook and aircraft's responses to catching and breaking this weak line.

The first mat landing was performed on June 2, 1954, but was unsuccessful. The aircraft, S/N 51-1225, was piloted by a Martin Aircraft test pilot. The tailhook missed the arresting cables and tore through the mat surface, tearing open three air cells. Apparently the test pilot was not aware that the F-84 had a tail-hook/airplane flap interconnect system that automatically retracted the flaps when the tail hook contacted the arresting cable, or that he had a manual override switch. The momentary contact between the tail hook and the mat was enough to cause the flaps to retract and the

aircraft to settle on the mat too quickly. Complicating the problem was the slow engine response to the pilot's full throttle command. The F-84 bounced off the mat, skidded across the lakebed, and was damaged beyond economical repair. The pilot received back injuries that grounded him for several months.

While the mat was being repaired, the launch phase of the program continued. All launches were completed by August 1954. Fourteen successful launches were made with the T-50 rocket and eight successful launches with the X-224-B1 rocket.

The second mat landing was performed on December 8, 1954. While considered successful, the pilot suffered a strained neck due to the high pitch rate of 62 degrees per second when the hook engaged the arresting cable. On the third try on December 11, the aircraft landed as expected. The F-84 caught the cable at 144 miles per hour, pitched down, hit the mat and bounced up into the air four or five feet, came to a stop essentially in mid-air, then hit the mat for a final time. Graphite was used to lubricate the mat on this third trial.

Although thirty mat landings had been planned, the ZELMAL program was canceled on December 16, 1954, after these three mat landings. It was felt that sufficient data had been obtained on the two successful landings to come to a conclusion and prepare a report. The biggest drawback with the mat landing was that the pilot's head was snapped around rather violently, experiencing vertical load factors of 5.4 to 6.1 Gs, and longitudinal load factors of 3.4 to 5.4 Gs. No consideration had been made for a pilot head restraint. The program completed 22 launches and three mat landings. The various test reports did not specify if any of the mat landings were performed following a zero length launch.

F-100 ZERO-LENGTH-LAUNCH

In 1957. military planners still saw the value in the zero-length-launch concept, but the idea now was to launch atomic bomb equipped fighters from roving trailer trucks for a retaliatory strike against the Soviet Union. This capability could present a credible deterrent to an enemy's first strike. In a nuclear

This F-100 is shown right after launch from its mobile launcher. (USAF Photo)

The booster was oriented in this manner so that the thrust line went through the center of gravity. (USAF Photo)

exchange, in which the winner would be determined within two, maybe three volleys of nuclear weapons, the cost of a few dozen, or even a few hundred fighters was insignificant. Thus, the whole concern for the landing became irrelevant...after the mission was accomplished, the pilot would fly to friendly territory and eject.

The F-100 was chosen for this effort. Weighing 35,000 pounds, more than twice the weight of the F-84, the F-100 would be armed with the T-63 nuclear warhead. The aircraft would be boosted to flying speed using a XM-34 solid rocket booster produced by the Astrodyne division of North American Aviation. The XM-34 could produce 130,000 pounds of thrust for four seconds. This would boost the F-100 to 400 feet and 275 miles per hour. The pilot would experience 3.5 to 4 Gs in the longitudinal direction.

As with the F-84 program, the design of the rocket mounting assembly was critical. With this much thrust, the rocket's thrust line had to be aligned exactly through the aircraft's center of gravity, or else severe nose down, nose up, or yaw motions could result that might be beyond the pilot's capability to counter. Five test launches using a steel and concrete mockup were performed. Fears were confirmed when the mockup did a backward flip and other extreme maneuvers during test launches.

Two F-100Ds, S/N 56-2904 and 56-2947, were provided to the program. They were specially modified for this test program during production and bailed to North American Aviation. Altogether, five dummy-mass launches, fourteen piloted launches by North American Aviation test pilots, and six launches with Air Force test pilots were performed. Test weights ranged from 34,414 to 37,178 pounds, and with rocket boosters intentionally misaligned.

Launch configurations included various combinations of 200 and 275 gallon drop tanks. Ironically, the F-100 could not be launched in a clean configuration because the lighter

weight would have resulted in a longitudinal acceleration of 5 Gs, which would have been excessive. The F-100 was launched at a 20 degree angle from a mobile trailer 40 feet long, 12 feet wide, and weighing 55,000 pounds.

The first manned launch was performed in March 1958. It was flown by a North American Aviation test pilot using aircraft #2904, and went perfectly. However, the post flight inspection revealed that the booster had struck the aft fuselage upon separation. Inspection of the flight data showed that the booster separated during a momentary negative G loading. A retainer was incorporated to prevent the booster from releasing under negative G conditions.

The second flight was performed on April 11, 1958, also using 2904. This time a problem arose. The rocket booster would not separate after it burned out. The F-100 could not land with the empty rocket hanging under the fuselage. The pilot flew for an hour and a half trying to get the casing loose, but nothing worked. Finally, the pilot had to eject. It was determined that the attachment bolts didn't sheer off as designed, but not due to a failure of any component. The reason was that although all components were manufactured within their own tolerances, a condition known as "tolerance accumulation" ultimately caused a critical component to be out of position. All manufactured components have a specified dimension and an allowable tolerance. Normally, some components will be slightly oversize, and some slightly undersize, but still within their specified tolerances. Tolerance accumulation results when too many tolerances are either slightly high or slightly low, so the overall dimension for the entire assembly may be bigger or smaller than expected. To correct the prob-

With its nose pointed skyward at about 30 degrees, this F-104 waits for its "Kick in the pants." (USAF Photo)

Prior to launch, this F-104 has a special gantry moved into position for pilot entry. (USAF Photo)

lem, the booster attachment system was redesigned using exploding bolts. The new system worked properly throughout the remainder of the test program.

Fourteen successful launches were performed between March and October 1958. Launches were performed in tail winds as high as 17.5 knots. Three North American test pilots and one Air Force test pilot performed the flights. They all felt that no special qualifications were needed, and any combat qualified F-100 pilot should be capable of performing the zero length launch. The only critical portion of the launch was during the first few seconds. Because the airspeed was increasing so rapidly, it was easy to over control if a stick input was made immediately following rocket ignition.

A public demonstration was flown at an air show at the 1958 USAF Fighter Weapons Meet at Nellis AFB. As a sign of his confidence in the system, the Air Force pilot did a roll immediately after the booster fell off.

By this point, both the feasibility and practicality of the concept were verified. No one questioned that it worked. However, operational misgivings arose. The Europeans were wary of mobile nuclear weapons that could be hauled all over their territory, not to mention concerns for sabotage and accidents. Providing security for so many dispersed nuclear weapons certainly would have been a problem. They also questioned the need for launching from forward bases, when fighters with drop tanks could launch from bases farther away. To counter these arguments, launching from indoor, hardened bunkers was considered. On August 19, 1959, an uneventful launch was performed from inside a fixed building at Holloman AFB. As a last demonstration, a night launch was performed, also with no problems.

Finally, there was nothing left to test. Everyone agreed that zero length launch was a doable capability with no showstoppers. Provision for zero length launch was added to over 100 F-100s, but still, the operational capability was never fielded.

F-104 Zero-Length-Launch
By the early 1960s, the German Luftwaffe became interested in vertical take-off aircraft. They sponsored ZEL flights on an F-104 powered by a Rocketdyne engine. Tests were conducted either during spring 1964 or 1966 at Edwards AFB (references vary). Again, the launches were successful, but the capability never was fielded. No other information about the program could be located.

12
Vertical Flight Maneuverability

There is one other mission application where vertical flight provides a considerable offensive advantage in combat. That occurs when the near-vertical attitude flight capability of a fighter aircraft could spell the difference between victory and defeat.

The effectiveness of a modern fighter depends greatly on the effectiveness of its weapons. With earlier air-to-air missile systems, it was necessary for the pilot to line up his aircraft with the enemy and squeeze off the missile. With newer and higher-technology systems, with their so-called off bore-sight capabilities, a missile can make an intercept even though the enemy plane is not directly in the line of fire.

With the aerial dogfights of the future, conflicts won't always occur as high-speed swirling battles, but at slower speeds at high vertical attitudes. The capability will provide both offensive and defensive abilities to the aircraft.

With this attribute, the fighter can force on-coming missiles into high-drag, energy consuming changes in direction which, in certain situations, could break seeker lock-ons and escape interception.

Another term which is closely associated with near straight-up maneuverability is the term agility, which is defined as the ability to be quick and unpredictable while maneuvering. Having this ability at high angles of attack, when the velocity is quickly tailing off, can also be an extremely positive factor in close-in air-to-air combat.

So how is this unique combat capability acquired? First, there is the use of aerodynamics with such devices as forward fuselage canards and main engine thrust vector control, both of which can aid in maintaining nose attitude control when pointed skyward. Also, certain wing configurations can contribute to aircraft control while at high flight angles.

This capability has been extensively investigated during the 1980s and 1990s, with a number of positive results. High angle of attack (AOA) flight, as attempted by the innovative X-29 research aircraft, proved to be very effective with its forward-swept wing configuration. Similar research was also accomplished by the Russian S-37 forward-sweep aircraft, which also demonstrated outstanding maneuverability at high flight angles. In addition to its forward sweep wing, the S-37 also demonstrated a thrust vector control capability with its propulsion system.

Improved handling qualities at high AOAs was also examined by a joint NASA-Navy program where a Grumman F-14A fighter was modified to investigate a new control system in a program called the Aileron-Rudder Interconnect (ARI). The concept included a spin-recovery parachute and two-position deployable canard surfaces and produced increased high AOA maneuverability.

Of course, all modern fighters (like this F/A-18) have the capability to fly straight up. But at this attitude, there is a lack of maneuverability. A number of attempts have been made to improve it in recent decades. (McDonnell-Douglas Photo)

The F-16CCV research plane examined maneuverability at high angles of attack. (USAF Photo)

This chapter will detail a number of the significant programs addressing the high flight angle situation and the different techniques used to achieve it.

These programs include the F-15 STOL/MTD program, the F-15 ACTIVE program, and the F-16 MATV program. Most of these test aircraft had the capability of maintaining pitch control only, while the ACTIVE and MATV programs, with their axi-symetrical nozzle configurations possess both pitch and yaw capabilities.

The so-called Vortex Flow Control (VFC) technique is another area of research, which was carried out on the X-29 for increased stabilization at high AOAs.

NASA, in addition to the services, also investigated high AOA flight with its F/18 HARV vehicle. This technique uses three paddles around engine exhaust nozzles to deflect engine thrust to maneuver and stabilize the aircraft. Serious high AOA research has also been carried out with a pair of advanced experimental planes, the International X-31 and the NASA/Boeing X-36.

So here are the hows and whys of this other aspect of vertical flight, which takes place once the aircraft is at altitude:

VISTA/MATV NF-16 VARIABLE STABILITY AIRCRAFT (U.S.A.)

System: System: Modified F-16D
Manufacturer: Lockheed Corporation, Calspan Corporation
Sponsor: USAF
Operating Period: 1993 - 1994
Mission: In-flight Simulation of high AOA flight stability
The Story:

VISTA

The Variable Stability In-Flight Simulator Test Aircraft, or VISTA, which was the base model for the MATV version to follow, was built as a new in-flight simulator that would have performance representative of the latest operational fighter aircraft. Built around a production F-16D airframe, the VISTA comprised production features from many different F-16 vari-

The Aileron-Rudder-Interconnect F-14 research plane was used to investigate high high angle-of-attack maneuverability. (US Navy Photo)

The F-16 VISTA was the next attempt at high-AOA maneuverabiity. (USAF Photo)

ants, plus a complex custom-designed variable stability system to allow the VISTA's flight characteristics to be changed.

The airframe design was a mix of different features already available from other production versions, including a large dorsal housing, heavy weight landing gear, and digital flight control computer. The gun, ammunition drum, and many unneeded defensive systems were deleted. A larger capacity hydraulic pump and larger hydraulic lines were installed. A programmable center stick controlled by its own digital computer was installed in the front seat. The variable stability system design centered around three Rolm Hawk digital computers to determine the motions of the simulated aircraft and the necessary motions of the VISTA control surfaces to produce them.

Controls to access the computer, in order to change flight characteristics, and to engage the front seat controls, were installed in the rear seat. Extensive automatic safety monitoring would watch the VISTA's motions and instantly return control to the safety pilot in the back seat if a potentially dangerous situation was being approached. The VISTA made its first flight in April 1992.

But, as fate would have it, another Wright Laboratory program was gearing up which needed a F-16D for its test vehicle for the identical time period that VISTA would be on hold. General Electric had already developed a nozzle that could deflect the engine thrust up to seventeen degrees off the thrust axis in any direction (this could produce up to 4,000 pounds of force up, down, or sideways on the tail of the aircraft). It was anticipated that the combination of thrust vectoring and control surface motions could greatly improve the F-16's maneuverability at high AOAs.

MATV

This program, called MATV for Multi-Axis Thrust Vectoring, was to demonstrate the improved combat effectiveness of a F-16 equipped with this nozzle. Thus, the Wright Laboratory "loaned" the VISTA aircraft to the MATV program to accomplish the testing.

Actual modifications to the VISTA aircraft began in Summer 1992. From the outside, the aircraft looked about the same, except for the addition of a spin chute and the painting of the MATV logo on the tail. The vectoring nozzle was indistinguishable from a standard one except to the trained eye.

Internally, the modifications were more extensive, with most of the VISTA-unique systems removed so that from a functional standpoint, the airframe was essentially a standard F-16D. The software in the digital flight control computer was modified so that when the thrust vectoring mode was selected, the nozzle moved with the elevator and rudder. No special commands by the pilot were needed.

The VISTA made its first MATV flight in July 1993 and continued through March 1994 when it made its 95th and final MATV flight. The first phase of these flights investigated using thrust vectoring to maneuver the aircraft in the "post

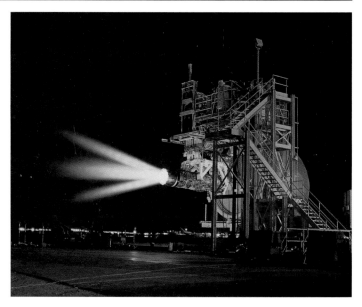

The MATV version of the VISTA demonstrated a triple-axis propulsion system, shown here in test. (General Electric Photo)

stall" flight regime. What this means is that the aircraft is actively maneuvered even through the wing was stalled!

Traditionally, aircraft are maneuvered up to the stall, but not beyond because of the possibility of entering a spin. Using thrust vectoring to control the pitch and yaw, the aircraft's nose can still be pointed as desired, even though the wing is stalled and the control surfaces may be ineffective. Thrust vectoring allows the pilot to point the nose wherever he wants without worrying about loosing control.

Sounds great in theory, but the combat effectiveness of this capability had to be demonstrated on an aircraft representative of a top-line fighter. Four specific goals of the MATV

The MATV test aircraft demonstrates its high AOA capabilities. (USAF Photo)

program were to 1) determine the flight envelope for thrust-vectoring maneuvers, 2) evaluate the close-in, combat utility of this enhanced maneuvering capability, 3) assess the aircraft's flying qualities while using thrust vectoring, and 4) examine the engine's operability at extreme attitudes and angular rates. A number of post-stall maneuvers were demonstrated.

• The "Cobra," a fast pitch-up to a straight-up attitude or beyond, then back to level flight at almost the same starting altitude. This maneuver could be used to get a quick shot at an overhead aircraft or even a missile shot at an aircraft approaching from behind.

• The "J-turn," a rapid pull up to vertical, followed by a maneuver until the nose is pointed down, then a sharp pull-up, resulting in a very quick 180 degree heading change.

• The "Hammer Head," in which the aircraft starts a loop. When vertical, the pilot makes a 180 degree post stall rotation about the pitch axis, resulting in a vertical, nose-down position. From there, the pilot completes the loop.

Tests conducted with the MATV quickly demonstrated its ability when pitted against standard F-16s. One F-16 pilot indicated that flying against the MATV was nearly impossible. No matter where the F-16 was located with respect to the MATV, the latter could quickly turn and point directly at the F-16. A definite tactical advantage to be sure!

The MATV program demonstrated a new flight regime for controlled, high-AOA flight. The vehicle greatly expanded the F-16 flight envelope, increasing its angle of attack capabilities from about 25 degrees to more than three times that figure at about 80 degrees. In fact, during the development stage, the MATV demonstrated maneuvering beyond 125 degrees without any pilot restriction.

An aircraft with such a high angular capability would have a significant advantage in any one-on-one or even one-on-two scenarios.

The MATV also proved itself effective at high AOAs at low airspeeds where the elevator and rudder were ineffective.

The system vividly demonstrated a real, tactical advantage for thrust vectoring with its completely integrated thrust vectoring system.

HARV HIGH ANGLE-OF-ATTACK TEST VEHICLE (U.S.A.)
System: Modified F/A-18
Manufacturer: McDonnell Douglas Corporation
Sponsor: NASA
Time Period: 1980s and 1990s time period
Mission: To demonstrate the capability for high angles-of-attack.
The Story:
Flying at high vertical attitudes, problems are created when the airflow around the aircraft becomes separated from the airfoils. At high angles-of-attack, the forces produced by the aerodynamic surfaces—including lift provided by the wings—are greatly reduced. This flight attitude often results in insuf-

ficient lift to maintain altitude or control of the aircraft.

The NASA HARV program produced technical data at high angles-of-attack to validate computer analysis and wind tunnel research. Successful validation of this data enabled engineers and designers to understand better the effectiveness of flight controls and airflow phenomena at high angles-of-attack. The goal of this ambitious program, simply stated, was to provide better maneuverability in future high performance aircraft and make them safer to fly.

The program began in 1987 with the first of three phases. The initial phase lasted for two and one-half years and consisted of 101 flights with the specially-modified F/A-18 HARV (High Angle of Attack Reserch Vehicle), which flew up to angles-of-attack approaching 55 degrees.

To measure the test plane's performance, visual studies of the airflow over various parts of the aircraft were performed. Special smoke was released through small ports just forward of the F-18's leading edge extensions near the nose, and photo- graphed as it followed airflow patterns around the aircraft.

Also photographed in the airflow were small pieces of yarn taped on the aircraft, along with an oil-based dye released onto the aircraft surfaces from 500 small orifices around the aircraft nose. Additional data obtained included air pressures recorded by sensors located in a 360-degree pattern around the nose and at other locations on the aircraft. All these measurements and flow visualization techniques helped engineers to correlate theoretical predictions with what was actually happening.

Phase two of the HARV program began in summer 1991, using a considerably-modified HARV vehicle for new test mission requirements.

Three spoon-shaped paddle-like vanes, made of inconel steel, were mounted on the airframe around each engine's exhaust. The devices deflected the jet exhaust to provide both pitch and yaw forces to enhance maneuverability at high

The F/A-18 HARV test aircraft was NASA's research effort at improvement of high-AOA capabilities. (NASA Photo)

The HARV demonstrated high angle-of-attack capabilities of up to 70 degrees. (NASA Photo)

angles-of-attack when the aerodynamic controls were unusable or less effective than desired.

Also, the engines had their external exhaust nozzles removed to shorten the distance by two feet to the vanes that had to be cantilevered. The unique thrust-vectoring modification precluded the HARV aircraft from being able to accomplish supersonic flight. The system added over two thousand pounds to the craft's weight.

The modifications enabled the HARV to be much more effective at higher angles-of-attack, up to near 70 degrees. Since the aircraft could hold the high AOA for longer time periods, much more data was available for collection.

The envelope expansion flights were completed in February 1992. Demonstrated capabilities included stable flight at 70 degrees, up 15 degrees from the previous high of 55 degrees, and rolling at high rates at 65 degrees angle of attack. This phase of the program was concluded in late 1993.

During early 1994, the final phase of this program was undertaken, with additional modifications of the HARV test aircraft being accomplished. Changes included the addition of movable strakes which were mounted on both sides of the nose to provide enhanced yaw control at high angles-of-attack. The four-foot-long, six-inch-wide strakes were hinged on one side and mounted flush to the forward sides of the fuselage.

At lower angles-of-attack, the strakes were folded flush against the aircraft skin, while they were extended at higher angles in order to interact with vortices generated along the nose. As a result, large side forces for enhanced control were produced. Early research indicated that the strakes could be as effective as rudders at the lower angles of attack.

Looking at the HARV F/A-18 aircraft itself, it is a modified version of a pre-production F/A-18 Hornet. The plane retained its standard GE F404 powerplants, each capable of 16,000

The STOL F-15 design in early wind tunnel testing. (McDonnell Douglas Photo)

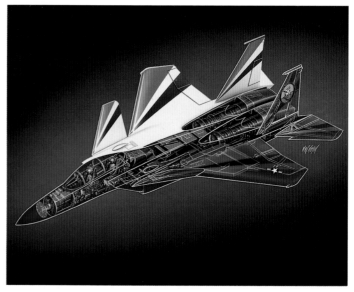

Internal layout of the F-15 ACTIVE test vehicle. (McDonnell Douglas Photo)

pounds of thrust in afterburner. Typical take-off weight for the HARV is about 39,000 pounds, with 10,000 pounds of internal fuel, thrust vectoring, and a unique spin recovery parachute system installed.

F-15 ACTIVE HIGH ANGLE-OF-ATTACK MANEUVER PROGRAM (U.S.A.)
System: Modified F-15B
Manufacturer: McDonnell Douglas
Sponsor: USAF
Operating Period: 1985 to mid-1990s
Mission: Development of short take-off and landing and high angle-of-attack maneuver capabilities
The Story:
The goal of the F-15 STOL/MTD (ACTIVE forerunner program) was to develop the capabilities for future fighters to land on wet, short, bomb-damaged runways. In addition, the testbed aircraft also demonstrated the capability to improve combat maneuverability with greatly increased capabilities at high angles of attack. There was no mistaking this testbed aircraft with its flashy red, white, and blue paint job.

The aircraft, a highly-modified F-15B Eagle, participated in a 1985-to-1991 test program to validate its unique Pratt & Whitney-built two-dimensional thrust vectoring/reversing fighter nozzle configuration. In addition to its unique engine nozzles, the STOL/MTD testbed was also distinguished by moveable control canards mounted on the forward fuselage. Other modifications included integrated flight and propulsion controls, rough-field landing gear, and an advanced cockpit which greatly reduced the pilot/vehicle workload, in spite of the complexity of the maneuvers to be performed.

The initial phase of the program began in September 1988 and recorded 43 flight tests with standard, circular nozzles. Then, the Air Force and contractor personnel installed and ground-tested the aircraft's thrust-vectoring, thrust-reversing nozzles which were designed and manufactured by Pratt and Whitney.

The first flight of the F-15 STOL/MTD with the new nozzles was May 10, 1989, from Lambert-St Louis International Airport. On June 16, the craft was ferried to Edwards Air Force Base, California. Shortly thereafter, the F-15 STOL/MTD demonstrated and evaluated the thrust-vectoring feature of the new nozzles coupled with the integrated flight/propulsion control system of the aircraft.

The STOL/MTD expanded its in-flight thrust reversing envelope to Mach l.6 at 40,000 feet. Assisted by thrust vectoring, the aircraft's pitch maneuverability at high angle-of-attack was demonstrated to be 110 percent better than an aircraft with only conventional controls. A flight control system modification in April 1991 lowered the landing distance to 1,370 feet, less than one-third that of a normal fighter.

For its accomplishments, the program was honored with the 1990 Aerospace Laureate Award for outstanding contributions to aerospace technology.

F-15 ACTIVE Program
Following its use in the STOL/MTD Program, this particular sophisticated testbed aircraft could have languished in storage for many years. It has happened with other testbeds in the past.

But there was another job to be performed, and this highly-modified Eagle was brought back to life after 22 months of

The F-15 STOL/MTD used advanced new flight technologies to improve high AOA maneuverability. Note the forward-mounted canards mounted on the upper part of the engine intakes. (USAF Photo)

storage. The effort was the so-called ACTIVE Program, for Advanced Control Technology for Integrated Vehicles, which was a joint NASA/USAF/Pratt & Whitney effort.

The purpose of the program was to investigate symmetric pitch/yaw vectoring nozzle technology with an even more-advanced nozzle that had been fitted to the F-15. This nozzle, similar in concept to the one installed on the VISTA F-16 testbed, could deflect thrust in any direction, not just up and down.

Testing indicated that the advanced nozzle technology could result in the use of nozzles for aircraft trim, thus reducing drag, both during maneuvering at high angles of attack and in cruise flight.

For the ACTIVE Program, it was necessary to make some structural modifications for the aircraft to accept the new engine nozzles and fly with them. The airframe was also reinforced to accept the increased yaw forces, and the fuselage fairings were changed so that the vectoring engine nozzles would fit where the original engines had been located.

X-29 VORTEX FLIGHT CONTROL (VFC) HIGH ANGLE-OF-ATTACK SYSTEM (U.S.A.)
System: Modified X-29 system
Manufacturer: Grumman Aircraft Corp.
Operating Period: 1992 time period
Sponsor: USAF
Mission: Maneuverability at high angles-of-attack
The Story:
In the future, the slightest advantage one pilot has over an adversary in an aerial dog fight could mean the difference between victory and defeat.

A Vortex Flight Control (VFC) system, using small nose-mounted nozzles to influence the natural airflow around the nose of the aircraft, could provide control at high AOAs. This innovative approach could provide a friendly pilot the capability to point and stabilize the aircraft's nose more quickly in a high-AOA condition in order to release an air-to-air weapon.

The capability to develop a VFC system to address this high AOA problem was investigated during the mid-1990s with a flight test program at Edwards Air Force Base using the X-29 system.

A research initiative of the USAF's Wright Laboratory, the program was a joint effort of Wright Laboratory, NASA Dryden, the Air Force Flight Test Center, and Grumman.

Lt Col William Gotcher, then X-29 program manager at Wright Laboratory's Flight Dynamics Directorate, explained, "We're breaking new ground in aviation history by generating directional control through influencing vortex control. This is the first time it's been done. There was earlier work done with reaction control jets on the F-104, but none involved vortex flow control. This is a new technology."

The development of such a VFC system could have important implications in the design of future fighter aircraft. The payoffs of this new VFC technology could result in better directional control at high AOAs, reduction in the size of the tail, or elimination of the tail completely.

The X-29 test set-up consisted of an on-board high pressure nitrogen system to power a pair of nose-mounted nozzles to influence the vortex flow over the body. Two nozzle sizes were investigated, with one-half and one-pound-per-second flow capabilities.

The VFC flight testing was preceded by wind tunnel testing, accomplished both at Grumman and Wright Laboratories. "It was through the Wright Lab testing that we came up with the slotted nozzle design that was used," Gotcher said.

"The wind tunnel testing showed that the VFC was a viable concept. With the flight testing, we found that the results were even better than we expected. In just 12 months we took VFC from wind tunnel testing to full-scale flight testing, a process which usually takes several years." Results showed that the on-board VFC system becomes effective at 15 degrees angle of attack, surpasses rudder power above 30 degrees, with the capability of demonstrating much higher angles.

Construction of the X-29 forward-sweep test aircraft. (Grumman Photo)

Gotcher emphasized, though, that in an operational configuration, the nitrogen system would not be used: "Such a system would probably use engine bleed air to power the nozzles. That's something which wouldn't be that difficult to accomplish. Down the road, we could see an operational fighter modified with such a system for testing."

It was determined that such a system in an operational aircraft would probably be incorporated into the flight control system, which would respond to a command and activate the VFC system, demanding enough nozzle blowing to accomplish the desired maneuver.

X-31 HIGH ANGLE-OF-ATTACK RESEARCH AIRCRAFT (U.S.A./GERMANY)

System: Built-from-scratch research aircraft with many parts and pieces from F-16 and F-18 systems
Manufacturer: Rockwell International/Messerschmitt (MBB)
Operating Period: 1983-to-mid-1990s
Sponsor: DARPA, U.S. Naval Air Systems Command (both USA) and German Federal Ministry of Defense
Mission: A system designed to break the stall barrier to accomplish effective close-in combat at high angles of attack
The Story:
This international program began in the early 1980s when both Rockwell International and MBB were studying the capability for fighter aircraft to be able to fight at normally near vertical attitudes. The program was initially included as a part of a European program. Eventually, the two companies would come together in an Enhanced Fighter Maneuverability (EFM) program which would enable the building of the two X-31 prototypes.

There would be two X-31s built, with advanced composite materials being used extensively in the wing and thrust-deflecting vanes. The X-31 featured a low-mounted cranked-delta-wing design with a 56 degree inboard leading edge sweep. Large trailing and leading edge flaps emphasize the control function constituted by the wing. There are also a pair of small canard fins located just forward and below the canopy. A single vertical fin constitutes the tail assembly.

The X-31 is powered by a GE F404 turbofan which is fed by a F-16 style lower intake. A set of paddles directed aft of the engine enabled a considerable thrust vectoring capability, enabling the test fighters to accomplish extremely tight maneuvers.

The plane grosses out at about 14,000 pounds with an empty weight of just over five tons, with a maximum altitude capability of over 50,000 feet. The X-31 made its first flight in late 1990, and the two aircraft would make over 500 test flights through 1995. Unfortunately, one of the prototypes would crash during a 1995 test.

The X-31 proved capable of achieving high vertical attitudes and remaining in control, thus producing a situation that presents an advantage against an adversary in air-to-air fighter dogfights.

Later in the program, the X-31 demonstrated an amazing capability by executing the so-called "Herbst Manuever." The maneuver consists of entering at a high speed (Mach .5 or better), then decelerating rapidly while increasing its vertical attitude. At this point, the plane exceeds conventional aerodynamic limits and needs thrust vectoring for control. Suddenly, the X-31 assumes a new flight direction, then lowers its nose and again accelerates to high speed while flying in the opposite direction. Certainly a maneuver that would have any enemy fighter wondering what had happened.

Then, in 1995, the program ended, apparently. But in 1998, the remaining X-31 came out of retirement for a new reserach program known as VECTOR, which was to examine extremely short take-off and landing along with the concept of tailless flight.

X-29 model is tested at the Air Force's AEDC wind tunnel facility. (USAF Photo)

X-36 UNMANNED LOW-SPEED, HIGH-ANGLE-OF-ATTACK VEHICLE (U.S.A.)

System: Built-from-scratch prototype
Manufacturer: Boeing
Operating Period: 1990s time period
Sponsor: NASA
Mission: To evaluate aircraft agility at high angles of attack and low speed with a remotely-piloted, manless, and tailless vehicle.

The Story:

During the 1990s, Boeing and NASA have combined for an interesting program addressing fighter-type aircraft at low speeds and high angle-of-attack flight, conditions where the airflow over the wings will not normally permit maneuvering.

The program, which has used minimal expenditures compared to other programs, uses a pair of 28 percent scale tailless vehicles to accomplish the experimentation. The unique design of the vehicles employs split drag rudders and thrust vectoring for control and directional stability.

The X-36 has a wheeled landing gear and has the capability to take-off and land under its own power. The model also carries a parachute so that recovery can be accomplished should control be lost.

Grumman X-29 testbed in flight. (Grumman Photo)

13
VTOL Concepts that Did Not Make It

Granted, through the years there were a multitude of VTOL vehicles that were researched and developed. Some made it to a prototype stage, fewer made it to a flying status, and just a chosen few made it to a series production status.

Hidden in old file cabinets and engineers' lower desk drawers were a number of long-forgotten or little-publicized VTOL concepts that introduced some new concepts. Some got interest from the military or commercial markets, while others languished in obscurity. Needless to say, there were some interesting innovations devised, with some reaching the level of almost unbelievable. But since this book addresses the subject of VTOL aircraft, some of these concepts deserve a final look.

Wingless Compound Helicopter (Sikorsky)

During the 1960s, the Sikorsky Company designed a wingless compound helicopter. The concept featured all the basic helicopter features, with the addition of either auxiliary propellers or jet engines to provide thrust for forward propulsion. Not surprisingly, the company indicated that all of the lift was provided by the main rotor.

Stowed Rotor Compound VTOL Vehicle (Lockheed-California)

Falling into the Compound Vehicle category, the Stowed Rotor concept vehicle would have presented some significant technical problems had it ever reached the metal-cutting stage.

The idea behind this paper effort was to convert a rotorcraft into a fixed-wing aircraft that could cruise at more than 500 miles per hour. Following a pure-vertical take-off, the craft would stop its rotor blades once horizontal flight had been achieved, and then fold them back atop the fuselage in a stacked configuration. It should also be noted that there was also a conventional helicopter-type rotor mounted on the horizontal member of the tail.

The achievement of the horizontal flight phase would be accomplished by a pair of wing pylon-mounted turbojet engines.

Rotor/Wing Vehicle (Hughes)

This Hughes concept devised an aircraft that could combine the speed of a fixed-wing jet with the hovering efficiency of a helicopter.

Long gone and forgotten, the Avro Avrocar languishes in a park at Fort Eustis, Virginia. (Air Force Museum Photo)

The flowing saucer-appearing Avro Avrocar was an experimental VTOL research vehicle using the Vectored Thrust technique. (USAF Photo)

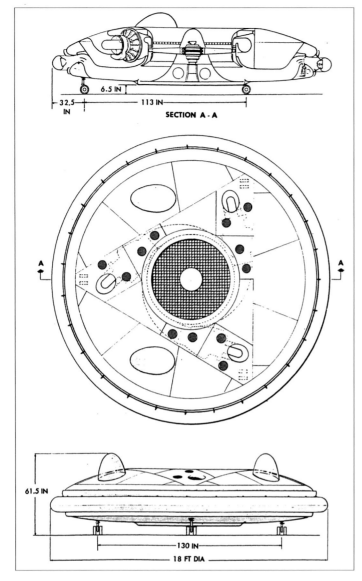

SECTION A - A

6.5 IN
32.5 IN
113 IN

61.5 IN
130 IN
18 FT DIA

A three-view drawing of the Avro Avrocar. (USAF Drawing)

The flight concept involved a vertical lift-off with the large overhead triple-ended wing acting as a rotor, with a hot cycle pneumatic drive system sending the turbojet exhaust through its rotor-tip jets.

Then, at about 90 miles per hour, the exhaust was to be diverted from the rotor to the tailpipe to five conventional jet thrusters for forward propulsion. The aircraft continued to accelerate with the upper wing free-wheeling until about 170 miles per hour was reached. The wing would then stop in position, shown by the Hughes illustration, where it would then function as a normal fixed aerodynamic surface.

U.S. Army Aerial Jeep Concepts (Aerophysics, Chrysler, Piaseski)

It was a VTOL without doubt, but the Aerial Jeep concept was also an Army Jeep-type vehicle. Combined together, it would have resulted in an effective vehicle for a number of purposes for a ground infantry-type unit.

A number of different companies were involved with some radically-different concepts. The Aerophysics Development Corporation featured a center platform with four large ducted fans at each corner. There was side-by-side seating for the two man crew. Originally carrying landing skids, the final concept was to have had standard wheels.

A Chrysler concept contained two large fans mounted within the flat platform. Between the fans were the crew compartment and a recoilless rifle. The Piasecki Helicopter had two fans in tandem. Unlike the other concepts, this concept carried a pair of fins on the back of the vehicle. The two-man crew sat between the fans behind a pair of wind deflectors.

The Convoplane (Goodyear)

If you think this concept looks like a flying saucer, that would be a correct assumption. The unique VTOL carried a pair of counter-rotating fans within the saucer fuselage which were to provide the vertical lift for the machine. The slipstream was

McDonnell Douglas Tilt-Thrust VTOL transport. (McDonnell Douglas Photo)

Vertol Retractable Rotor transport concept. (Vertol PHoto)

Sikorsky wingless compound helicopter. (Sikorsky Photo)

Artist's concept of Piasecki's flying jeep. (Piasecki Drawing)

moved through top and bottom louvers. Undoubltedly, had it ever been built, it would have been highly maneuverable. The single crewman was housed in a small cockpit which protruded from the leading edge of the front of the craft. Stabilization was aided by a tail of sorts mounted on the rear of the vehicle. The large horizontal tail was supported by three vertical members.

The company felt that the aircraft would have the capability to hover with high efficiency, low downwash velocity, low noise level, and the easy control of a helicopter, but with the capability to cruise at 500 miles per hour. It was also estimated that such a vehicle would have a high payload-to-empty weight because of its lightweight hot cycle drive.

K-25 Combination Tilt Wing/Tilt Rotor VTOL (Kellett)
An interesting joint concept vehicle was proposed that was to be a compromise between Tilt Wing and Tilt Rotor. The technique was accomplished by employing collective control and longitudinal cyclic control in each rotor with a ten percent flapping hinge offset.

By tilting the wing with the rotor, the stiffness of the support structure for the rotor could be kept fairly constant, thereby minimizing mechanical instability problems. Another feature gained was the reduced stress level on the blades through

McDonnell Douglas Tilt Pro concept vehicle using propellers at front and rear of each nacelle. (McDonnell Douglas)

This VTOL concept looked like a flying saucer, a design that was proposed by Goodyear in a contract with the Navy. (Goodyear Drawing)

Artist's concept of a flying saucer VTOL.

This concept VTOL transport featured a Tilt Rotor plus forward thrust.

A transport concept vehicle using a Tilt Wing.

transition. This configuration also had the ability to enter a partial transition without the necessity of completing the maneuver.

Model 70 Deflected-Slipstream VTOL (Fowler)
During the mid-1960s, a novel VTOL concept was introduced by Harlan Fowler, famous inventor of the Fowler Flap. The concept involved the use of a novel drag regulator, which, when combined with shrouded propellers and triple-slotted flaps, the device was to complete a Deflected Slipstream system which Fowler indicated could produce turning angles of nearly 90 degrees.

The purpose of the drag regulator was to balance the thrust of the propellers, and reduced the wheel braking power required. The device was to be stored in a steamlined, spanwise strut housing, and hauled down like a window shade during VTOL operations.

Fowler proposed the concept in a Model 70 design, a cargo or passenger-carrying craft with a gross weight of almost 11 tons. It was to be powered by two GE T64 turboshaft engines driving four shrouded 6.5 feet diameter propellers through interconnecting shafts. A transmission system was to permit either engine to drive all four propellers.

CL-379 Deflected Slipstream VTOL (Lockheed)
This concept obtained vertical lift by increasing the wing incidence angle to about 20 degrees and using the flap system to give moderate slipstream deflection. In hovering flight, roll was to be controlled by differential pitch of the propellers, or spoilers in the flap system. Pitch was controlled by motion of

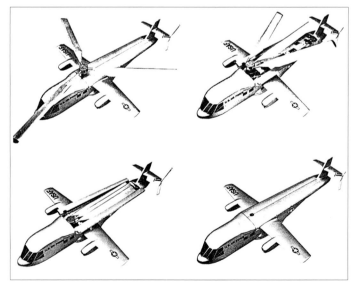

This stowed-rotor design of Lockheed was a design to convert a rotor craft into a fixed wing aircraft with a 500mph capability. (Lockheed Drawing)

This Army concept vehicle used a three-bladed rotor for conventional flight with one blade against the fuselage with the other two serving as wings. (US Army Photo)

The Lockheed CL-379 light fighter. (Lockheed Photo)

the flap sections augmented by the opposite motion of the leading edge slat, while yaw is controlled by the differential motion of the flap sections. A Lycoming T53 960 horsepower turboprop was the planned powerplant.

C-6A V/STOL Deflected Thrust Transport (Boeing, North American)

This concept VTOL transport was the direct follow-on of the LTV XC-142. The mid-1960s program, which would never reach the metal cutting stage, proposed the use of several VTOL concepts, including Deflected Thrust engines located in pods under the wing roots, along with consideration of the Tilt-Wing and Compound concepts.

Both Boeing and North American received detailed study contracts in 1965 for the C-6A.

Model D-188 Deflected Thrust VTOL (Bell)

The Bell D-188 was to be a high altitude VTOL fighter for both land and shipboard use with a speed capability of Mach 2. The wings were high-mounted with negative dihedral and an unswept tapered planform. Propulsion was a pair of J79 engines mounted in the fuselage. Between the engines, and on the centerline of the fuselage, was a group of vertically-mounted J85 engines which was called the "Vertipack."

Each main engine had a diverter valve and a blocking valve which was used to direct the engine gases rearward for forward propulsion, or downward for direct vertical lift. The valves were placed directly to the rear of the engines and

North American concept vehicle for the USAF C-6A competition. The design features deflected thrust engines in pods under the wing roots. (North American Drawing)

Sikorsky stowed-rotor concept transport. (Sikorsky Drawing)

Sikorsky stowed-rotor concept transport. (Sikorsky Drawing)

The Bell D-188A Tilt Engine concept fighter. (Bell Drawing)

Drawing of the BAe P.103 Tilt-Engnine fighter. (BAe Drawing)

forward of the afterburners. The afterburners were used only in forward flight. The vertipack engines produced only vertical lift, but were capable of inclination from the vertical to produce a horizontal thrust component when required.

VC-400 Tilt-Wing VTOL (VFW-West Germany)
During the 1960s, a number of VTOL programs were carried out by the West Germany aerospace industry. One of the most interesting was the VFW VC-400, which proposed two Tilt-Wing installations for VTOL flight.

Both tilt wings were located high on the fuselage, with the front wing positioned directly behind the cockpit. The rear wing was placed in front of the tail assembly. Turboprop engines were located on each wingtip—a most-interesting concept to be sure.

Trailing-Rotor Convertiplane (Bell)
In the 1960s time period, Bell proposed a unique Compound concept VTOL. The concept used two GE CF-700 turbofans mounted in underwing pods to produce forward thrust, and additionally to power turbines for the rotor system in the hovering mode.

The rotor system mounted two 38-feet diameter three-bladed propellers mounted in pods on the wing tips. The rotor pylons were to be vertial for take-off, then rotate to horizontal for forward flight. The rotors then would fold together and would be stored in an aerodynamic shape of the pod.

The company hoped for a ceiling of 6,000 feet with an optimistic maximum speed capability of 500 miles per hour.

P.103 Tilt-Engine VTOL Fighter (BAe)
Also in the 1960s, the British BAe Company looked at a high-performance Tilt-Engine fighter with the P.103 designation. The concept considered the mounting of the engines in the middle and forward portion of the wing. The craft was unique

in that there was no horizontal tail, and the plane also carried a large fuselage-mounted canard wing for added maneuverability.

Model 39 Tail-Sitter VTOL (Temco)
The Model 39 was to be a single-place, fighter-class VTOL aircraft with low delta wings and both upper and lower vertical tail surfaces. A four-point landing gear was in place for its tail-sitting position. Performance estimate for the plane, which was to be powered by an Allison J-71 turbojet, was an airspeed of Mach 1.6 at 35,000 feet.

Flight control was conventional, with the stick used for pitch and roll control for level flight. For low speed and hover-

Concept by Short Brothers for the PD 25 ground attack aircraft. (Short Brothers Photo)

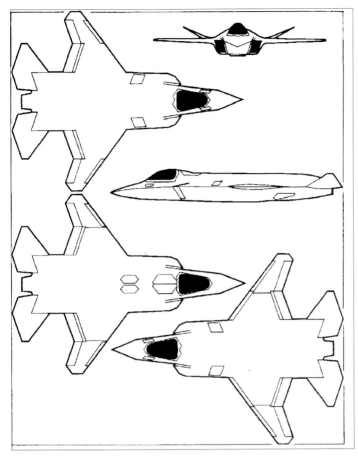

Losing McDonnell Douglas JSF fighter proposal. (McDonnell Douglas Drawing)

ing flight, the later motion of the stick was to control yaw. Roll was controlled by twisting the stick.

Flight control of the aircraft was accomplished with conventional aerodynamic surfaces. For low speed and hovering flight, a series of jet reaction controls were used. The rudder system was to be used only for directional control in normal level flight.

Avro Car (USAF)

An early concept program was the turbojet-powered Avro Car, a program which was taken over by the USAF after the Canadian government had investigated the program. The flying saucer project was based on model studies of a true disc configuration. Studies showed that flight control of the vehicle could be achieved through directional control of the jet exhaust issuing from a gap between the upper and lower disc surfaces.

The exhaust was to be directed up, down, or laterally from the perimeter by control of the exhaust gap dimensions to produce the Coanda effect. With one fixed gap dimension, the flow would be straight out to produce lateral thrust for transitional flight. With the gap precisely altered to another dimensional value, the exhaust tream was to bend back or up for a vertical thrust effect.

For vertical take-off, the top portion of the gap would be closed and alteration of the gap opening accomplished for downward thrust flow. For transitional flight, the gap, altered for straight-out jet flow and closed on a perimeter portion of the saucer, applied thrust on the opposite side of the perimeter of the vehicle.

Joint Strike Fighter (JSF) Losing Proposal (McDonnell Douglas)

Augmented by Northrop and British Aerospace, McDonnell Douglas in 1995 proposed a configuration for a Short Take-off, Vertical Landing configuration. The winning candidates in the competition would be concepts submitted by Boeing and Lockheed. (See Chapter 14 for details) The McDonnell Douglas proposal was terminated in November 1996.

As was the case with the winning proposals, there were to be three different versions of the fighter, i.e. USAF, Marine, and Navy shipboard models. The configuration was an extremely flat vhiecle, with the twin tail surfaces bent at about 30 degrees to the horizontal. The engine intakes were located at the aft end of the canopy in a F-22 manner. The swept wings were located far back on the mid-fuselage. The configuration appeared to be very stealthy.

14

Almost VTOL Aircraft Systems

Through the years, there have been a number of unique aircraft developments that have used the VTOL concepts enumerated in the preceding chapters.

The only difference with these systems is the fact that there is *not* the capability to accomplish the VTOL maneuver. Now, granted, the vehicles are greatly augmented with the VTOL technology, and their capabilities exceed a conventional aircraft.

It is for that reason that this chapter is included. The systems addressed include the pair of candidates for the Joint Strike Fighter (JSF) Competition. The fighters each have three versions—STOVL (Short Take-off/Vertical Landing), CTOL (Conventional Takeoff/Vertical Landing, and CV (Carrier Capable). Not one concept carries the desired VTOL connotation.

But even still, the two vehicles both use one of the aforementioned VTOL techniques; the Boeing X-32 uses the Vectored Thrust technique, while the Lockheed X-35 uses Lift Fan.

The so-called Credible Sport YMC-130H was a unique modification to the Lockheed C-130 transport. The plane was designed for a rescue mission in Iran in 1980. The plane, in addition to a number of other modifications, used a rocket assist system as the main assist for a STOL capability.

A pair of advanced transports, the Boeing YC-14 and the McDonnell Douglas YC-15, used the Deflected Slipstream concept to accomplish signficant STOL capabilities. The application of the technique, however, was differently applied with each model; the YC-14 used Upper Surface Blowing while the YC-15 acquired increased lift from pushing the thrust stream against the underside of the wing surface. Both produced the same effect.

BOEING X-32 JOINT STRIKE FIGHTER (JSF) VECTORED-THRUST FIGHTER (U.S.A.)

System: Built-from-scratch prototype
Manufacturer: Boeing
Sponsors: USAF, USN, USAF, RAF
Operating Period: Mid-1990s to present
Mission: One of two prototypes competing for the Joint Strike Fighter (JSF) contract. Winner of the competition will produce a light-weight, highly-maneuverable, VTOL fighter to be used by USAF, USN, and USMC, along with an expected large number of foreign customers. The Boeing concept will use a Vectored Thrust VTOL concept, but will not be a pure VTOL aircraft, since none of the three versions can perform both a vertical take-off and landing.

The Boeing X-32 is shown in its Marine version. (Boeing Drawing)

The Boeing X-32 in its Navy version. (Boeing Drawing)

This artist's concept shows a Royal Navy JSF taking off from carrier with a ski-jump ramp. (Boeing Photo)

The Story:

The requirements for the Joint Strike Fighter were rigorous. The goals of the program include the development of a highly-maneuverable, vertical-capability system that is capable of accomplishing three VTOL-type missions. The program actually began in 1995 under the JAST (Joint Advanced Strike Technology) program.

There are the STOVL (Short Takeoff/Vertical Landing) (USMC), CTOL (Conventional Takeoff/Vertical Landing) (Air Force), and CV (Carrier Capable) (USN) modes, all of which can be converted from the same model. All versions will have identical cockpits and avionics systems. In fact, Boeing indicated that there would be 70 percent commonality for all service variants, "Significantly reducing manufacturing, support, and training costs."

During the early 1999 time period, the flyaway costs of each of the three versions were $30-$35 million for the STOVL version, $28 million for the CTOL version, and $31-$38 million for the carrier capable type. The Boeing concept will carry more than 15,000 pounds of internal fuel, no matter which variant, plus from 14,000-18,000 pounds of payload.

Never before in fighter history has so much been expected of a new system. The JSF, be it the Boeing X-32 or the Lockheed X-35, will conceivably replace the USAF F-16 and A-10, the Navy F-14, early models of the F/A-18, and the USMC Harrier.

With the Marine STOVL version, there is a huge requirement for intake air. As such, this version of the X-32 has a gaping intake that can crank open to enable the gulping in of huge amounts of air.

The inlet duct measures five feet across at its widest point and is nine feet long, involving a complex curving shape. Boeing engineers explained that it was an extremely difficult shape to fabricate. In the Marine mode, the engine closes its vectorable cruise nozzle and opens a pair of lift nozzles at the center of gravity of the craft for vertical landings. The engine provides enough thrust for vertical flight. This version will carry a smaller weapons load than the other two versions.

In the CTOL Air Force mode, the expandable intake required for the STOVL mode is eliminated and a conventional take-off is accomplished. The simpler CTOL mission allows greater fuel to be carried, thus increasing the range of the version.

The Navy CV shipboard version will have a vortex-generating fence that can pump up on the upper wing surface during carrier approaches to assist in maintaining a high angle-of-attack.

Like the competing X-35, the X-32 uses the Pratt & Whitney F119 engine. Research on the performance of the engine has shown that increases could be made to even further enhance the capabilities of the JSF, no matter which concept was selected. Depending on the particular mission version of the X-32, there would be a particular version of the F119.

This engine is the most powerful fighter engine ever designed, producing its top thrust rating without the aid of an afterburner. The engine is installed in the forward fuselage and connected to its thrust vectoring nozzle by a straight duct.

Air Force version of the X-32. (Boeing Drawing)

Head-on view of Lockheed X-35 showing twin intakes. (Lockheed Drawing)

With the X-32 propulsion concept, there are two downward "posts" of thrust in front and a vectoring nozzle for the back. The model also uses under-wing nozzles for roll control, which are not directly connected to the engine. There are also small nozzles in the rear for yaw control.

Supportability is a key goal of the engine development program, with 60 percent fewer transports needed to deploy a wing of the aircraft and with support costs being about 60 percent lower. The overall configuration has a look that is completely different from any fighter ever designed. First, there is a blended delta wing, dual outward-cantilevered vertical tails, a large chine inlet, and thrust vectoring and side-fuselage weapons bays. The wing is a delta shape blending into the fuselage.

The large delta wing is extremely thick, allowing it to serve as an efficient fuel carrier and eliminating the need for external fuel tanks. The wings are fixed, but their short length will allow carrier storage.

From a production point-of-view, the Boeing JSF midfuselage section will be built in three sections. Additionally, the forward nose section can be constructed in single and two-seat versions, although in the late 1990s, no twin-seat versions had yet been confirmed.

Comparison of Lockheed JSF with F-16. (Lockheed Drawing)

Navy version of the Lockheed JSF. (Lockheed Drawing)

Dimensionally, the X-32 has an approximate wing span of 36 feet, with a fuselage bottom-to-vertical tail top of 15 feet. The overall length of the plane is only 45 feet. Maximum lift-off weight is in the 50,000-pound category, with an empty weight of only 22,000 pounds.

Like the X-35, with which it would be competing, two X-32s would be built for the flight test program. During that test phase, the planes will demonstrate all three modes of operation.

It should be emphasized that these first models were not actually full-up prototypes, but actually just the engine and the outer shape lines, filled with off-the-shelf components. The models will demonstrate low-speed handling characteristics when approaching a carrier for landing.

The discussion of whether the JSF would carry a gun was still being debated in the late 1990s. The Air Force felt strongly about having one, while the other services weren't that definite about having to carry along the extra weight.

LOCKHEED MARTIN X-35 JOINT STRIKE FIGHTER (JSF) LIFT FAN FIGHTER (U.S.A.)

System: Built-from-scratch prototype
Manufacturer: Lockheed Martin, Northrop, British Aerospace
Sponsors: USAF, USN, USMC, RAF
Operating Period: Mid-1990s to present
Mission: One of the prototypes competing for the Joint Strike Fighter (JSF) contract. Winner of the competition will produce a lightweight, highly-maneuverable near-VTOL fighter to be used by USAF, USN, USMC, RAF, and probably numerous other foreign customers. The Lockheed concept will utilize a Lift-Fan concept, but it is not a pure VTOL aircraft since none of the three versions is designed to perform that mission.

The Story:
To reiterate, the requirements for the Joint Strike Fighter were rigorous. The goals of the program included the development of the system to accomplish three distinct missions from three unique configurations. These included the STOVL (Short Takeoff/Vertical Landing) for the USMC, CTOL (Conventional Takeoff/Vertical Landing) for USAF, and CV (Carrier Capable) for the USN. All versions, however, carry the same basic airframes.

Like the Boeing X-32 entry, the X-35 versions are practically identical externally out to the wing-fold line. The only differences occur with the Navy CV version carrying larger leading and trailing edge flaps and larger horizontal stabilizers.

Unlike the Vectored Thrust concept of the Boeing Joint Strike Fighter configuration, the Lockheed Martin version would use a Lift-Fan system to achieve near vertical flight characteristics. The Lockheed JSF is based on an engine with an axisymmetric nozzle, similar to that used on the F-15 fighter. Pratt & Whitney has designed the F119 engine with easily-removable components to provide the lift and descent capabilities.

The STOVL version of the Lockheed JSF uses an Allison-developed lift fan which is powered by the main powerplant. The engine is connected to the fan via shafts and a reduction gearbox. The company found the Lift-Fan concept proved to be the best for their particular JSF concept for three reasons: the concept can lift more weight into the air for a given amount of thrust; it can provide a more benign ground impact of the downward air flow; and finally, it would allow the model to have smaller foward-facing air intakes, thus providing the design with a smaller frontal area, making supersonic flight easier to achieve.

Once the design is operating supersonically, the doors on the top of the fuselage open and suck in ambient air and augment it as it flows through the fuselage. The lift fan provides the air flow needed for hovering flight.

USMC version of the Lockheed JSF. (Lockheed Drawing)

Air Force version of Lockheed JSF. (Lockheed Drawing)

The Boeing YC-14 sees the light of day for the first time. (Boeing Photo)

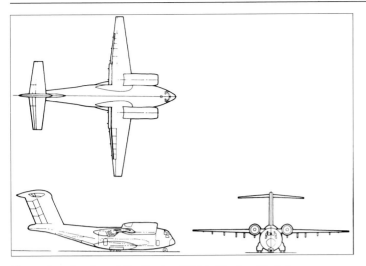

Three-view drawing of the Boeing YC-14. (Boeing Photo)

The rear-flap supports on the rear wings of the YC-14 are very visible in this photo. (Boeing Photo)

Further, the downward flow of the ambient air blocks any forward flow of the hot gas. And it will provide enough lift under the forward portion of the vehicle to balance the thrust from the air diverted downward from the hot-air nozzle at the tail of the aircraft.

All three versions of the X-35 utilize identical weapons bays. However, in the CV and CTOL versions, the lift hardware is removed and additional fuel substituted, making them more attractive for USAF and Navy use.

The Lockheed Skunk Works is assembling the two initial prototypes with flight tests expected to start in the first year of the new millennium. The prime contractor has indicated that the STOVL version causes the greatest challenge with concerns over propulsion and control. The CV version is next with its handling and control aspects. The simpler CTOL USAF variant causes the fewest worries.

YC-14 DEFLECTED SLIPSTREAM STOL TRANSPORT (U.S.A.)

System: Built-from-scratch system
Manufacturer: Boeing
Sponsor: USAF
Operating Period: Mid-1970s
Mission: To prove, along with the YC-15, the concept of Deflected Slipstream on the development of a STOL transport.
The Story:
Initially, it must be stated that the YC-14 was *not* a VTOL project, however, the use of the Deflected Slipstream concept makes it worth examining.

YC-I4 was the name that this research craft was best known by, one of two entrants in the so-called Advanced Medium STOL Transport (AMST) program of the 1970s. The other model in the program was the McDonnell-Douglas YC-15. The goal of both aircraft models was to design a large cargo aircraft with the capability to fly in and out of extremely short, semi-prepared fields.

Two YC-I4 models were produced under a $107 million contract with the Air Force. The YC-I4 was built to demonstrate the feasibility of a concept called upper surface blowing (USB), where thrust from the plane's two engines was used to blow air over the wing and flaps, thus creating powered lift.

The USB concept works on the situation that high speed air will follow the surface of both a wing and its accompanying flap system if the curvature is properly designed. The wing and trailing edge flap system of the YC-I4 allowed the engine exhaust air to create vertical, powered lift.

Exhaust air from the engines was blown across the wing's upper surface and down the curved flaps, much the way water from a faucet follows the back of a spoon. That sheath of high-speed air was able to turn almost 90 degrees and be directed downward. The model, with this arrangement, was able to take off quickly and land slowly at steep angles of attack on short fields.

During take-off, the large trailing edge flaps are functioning. Also note the high positioning of the engines. (Boeing Photo)

The YC-14 was designed to carry a majority of the Army's heavy equipment. (Boeing Photo)

To enhance the USB concept, the YC-14 used a so-called "Supercritical" wing, which combined the best aspects of high-lift and low-speed performance. The wing construction eliminated complicated wing-body joints and allowed the use of single-section skin panels and stringers.

As a result, the YC-l4 could take off from fields several times shorter than standard aircraft of comparable size. The plane could land on short fields at speeds as low as 99 miles per hour. Other new technologies on the YC-l4 were an advanced flight control system and a high-floatation, lever-type landing gear. With its innovative advances, the YC-14 could literally stop in mid-air, make extremely high-angle landing approaches, and roll to a stop in less than 1200 feet. With its drive-off unloading capability, less than a 2,000 foot take-off

roll, and 6,000 feet-per-minute initial rate of climb, the advanced transport would have had excellent applications at escaping enemy fire.

The YC-14's upper surface blowing concept, surprisingly, is not new, actually being an outgrowth of work done by NASA, and earlier, from work done by a Belgian scientist, Henri Coanda, who investigated the concept prior to World War II.

Aside from powered flight, the USB system also realized several other advantages. One is ingestion avoidance. With the engines above the wing and high up on the aircraft, the problem of the engine ingesting a foreign object, like a rock, is reduced.

This was particularly true in relation to the thrust reversers—the braking system used the thrust of the engines to

The YC-15 engines were mounted much lower than the counter YC-14. (McDonnell Douglas Photo)

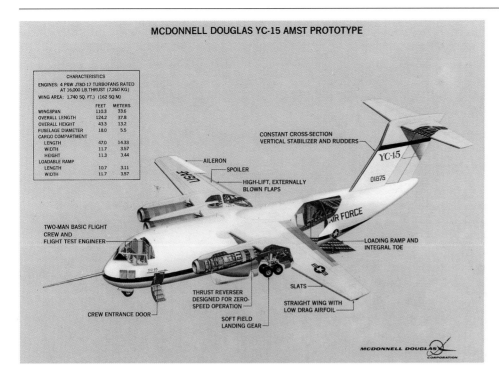

MCDONNELL DOUGLAS YC-15 AMST PROTOTYPE

CHARACTERISTICS
ENGINES: 4 P&W JT8D-17 TURBOFANS RATED
AT 16,000 LB.THRUST (7,260 KG)
WING AREA: 1,740 SQ. FT.) (162 SQ.M)

	FEET	METERS
WINGSPAN	110.3	33.6
OVERALL LENGTH	124.2	37.8
OVERALL HEIGHT	43.3	13.2
FUSELAGE DIAMETER	18.0	5.5
CARGO COMPARTMENT		
LENGTH	47.0	14.33
WIDTH	11.7	3.57
HEIGHT	11.3	3.44
LOADABLE RAMP		
LENGTH	10.7	3.11
WIDTH	11.7	3.57

CONSTANT CROSS-SECTION
VERTICAL STABILIZER AND RUDDERS

AILERON
SPOILER
HIGH-LIFT, EXTERNALLY
BLOWN FLAPS

TWO-MAN BASIC FLIGHT
CREW AND
FLIGHT TEST ENGINEER

LOADING RAMP AND
INTEGRAL TOE

THRUST REVERSER
DESIGNED FOR ZERO-
SPEED OPERATION

CREW ENTRANCE DOOR

SLATS

STRAIGHT WING WITH
LOW DRAG AIRFOIL

SOFT FIELD
LANDING GEAR

MCDONNELL DOUGLAS CORPORATION

The main parts and pieces of the YC-15.
(McDonnell Douglas Photo)

slow the aircraft when rolling. On some aircraft, thrust reversers pointed toward the ground and stirred up dust and small objects. On the YC-14, the thrust reversers were atop the engine and pointed upwards. This reduced the chance of anything blowing into the engines, along with pushing the aircraft into the ground and making wheel braking more effective.

Another advantage was noise. The wing shielded the engines from the ground, thus reducing the amount of engine noise that emanated downward.

The YC-l4 was powered by two General Electric CF6-50 fanjet engines, mounted near the fuselage above and at the forward edge of the wing. Each produced about 50,000 pounds of thrust. The YC-14 was about the same size as a Boeing 727 airliner, with a payload capability of about 27,000 pounds.

The engines were cantilevered far forward of the leading edge of the wing. They were also mounted quite close to the body, thus preventing a thrust misalignment problem during an engine-out situation.

No additional YC-l4s were produced after completion of the program, but many of the developed technologies would find their way into follow-on aircraft.

YC-15 DEFLECTED SLIPSTREAM STOL TRANSPORT (U.S.A.)

System: Built-from-scratch system
Manufacturer: McDonnell Douglas
Sponsor: USAF
Operating Period: Mid-1970s
Mission: Development of a deflected slipstream STOL
The Story:
The second part of the AMST program (the other being the Boeing YC-l4) was the McDonnell Douglas YC-15 transport. Two prototypes were constructed, but like its companion air-

The YC-15 with rear cargo door open and loading ramp lowered.
(McDonnell Douglas Photo)

YC-15 carrying camouflage paint scheme. (McDonnell Douglas Photo)

craft, the program would not continue beyond the prototype stage.

It should be noted, though, that the technology developed in the program would not go unused, much of which eventually being funneled into the C-17 advanced transport program of the 1990s. The YC-15 basically served as the basis for the company's winning C-17 proposal, with the C-17 looking remarkably like the YC-15.

There were basic differences between the look and operation of the two deflected slipstream STOL transports. Unlike the competitive YC-I4, the YC-15 used four engines, 16,000 pound thrust Pratt & Whitney JT8D-17 turbofans, instead of the YC-14's pair of larger engines.

The engines were mounted on the wings by means of pylons that pushed the engines far forward of the wing leading edge. In that location, the engine locations allowed the thrust to exhaust on the underside of the wing, directly into the large two-segment flaps. Recall that this is exactly the opposite situation that was accomplished with the YC-14, which pushed air over the top of the wing. The engines were fitted with special nozzles to mix the hot exhaust with cool air such that no special flap material was required. Inboard spoilers provided for the plane's lateral control and also aided in landing the craft.

With this interesting concept, the 33-foot long, double-segmented flaps were lowered directly in to the exhaust from the four engines. The location of those engines was such that the exhaust flow from each skimmed the underside of

the wings. A portion of the exhaust gas was then deflected downward by the flaps, creating lift. Another quantity passed through the wide spaces (Slots) between the flap segments and was turned downward by the Coanda effect to produce more lift.

McDonnell Douglas engineers estimated that this deflected thrust provided 20 percent of the YC-15's lift, and that super-circulation of air—air moving at an accelerated rate along the upper surface of the wing and downward along the

Four rocket boosters are fired during take-off and landing of the Credible Sport.

Shown are the forward-pointing rockets used for decelleration.

topside of the flaps—produced another 26 percent. The basic wing and flap, without power, generated the remaining 54 percent of the lift.

Another significant characteristic of the YC-15 was the increased cabin width, which permitted the YC-15 to carry up to 40 troops and six pallets simultaneously. Also, a large ramp in the rear of the aircraft facilitated vehicle loading, while a truck-bed level floor permitted straight-in access.

The YC-15 utilized the cockpit enclosure of the DC-10 tri-jet transport, modified with the addition of two downward-viewing windows to meet the aerial delivery requirement for additional visibility. The cockpit was designed for a crew of two, with room for a third crew member if desired. Reduced pilot workload was achieved through advanced technology, including use of a stability and control augmentation system, fly-by-wire direct lift control spoilers, and an advanced flight instrumentation system to supplement the full-powered controls.

With its advanced STOL characteristics reducing landing distances to less than two thousand feet for tactical missions, the YC-15 still was capable of cruising at more than 500 miles per hour on employment operations.

The military liked the capability because such speeds might have enabled military commands to complete 40 percent more sorties in a day's tactical operations. The YC-15's large fuselage, containing 67 percent more cargo space than the Air Force's largest tactical transport at the time, also made this prototype a very attractive system.

LOCKHEED YMC-130H "CREDIBLE SPORT" (U.S.A.)
System: YMC-130H
Manufacturer: Lockheed-Georgia Company
Sponsor: U.S. Air Force
Time Period: 1980
Mission: Super STOL to rescue American hostages in Iran
The Story:
Note—This program was very highly classified at the time it was performed. Although supposedly declassified, very little information was ever formally released. This story was compiled from various sources, including the Internet, and the validity cannot be ensured. As could be expected, there are gaps and contradictions in the information. Since virtually none of the information could be confirmed independently, conflicting information is included. Perhaps someday the full story will be confirmed. -SRM

After the failure of the hostage rescue mission in Iran on April 24, 1980, President Carter directed that a second res-

Credible Sport at Warner Robbins AFB Museum. Note the brackets that mounted rockets at one time. (Photo courtesy of Museum of Aviation)

cue effort be conducted before the presidential election in November. The Office of the Secretary of Defense (OSD) developed the plan, code named Operation Honey Badger. This time, helicopters would not be used. The rescue force would use a secretly-developed super STOL version of the C-130 Hercules transport that could land in a soccer field in Tehran. Delta Force would mount the rescue operation, but it was not clear if the commandos would arrive in the C-130s, or be infiltrated ahead of time, since the hostages possibly might have been scattered.

The mission was for these aircraft to take off from Eglin AFB in Florida, refuel in-flight on the way to Iran, and land in the Amjadien soccer stadium across the street from the U.S. Embassy in Tehran. After rescuing the hostages and departing, the aircraft were to land on an aircraft carrier in the Persian Gulf.

Naturally, there was concern over the hostages' locations, but by June 26, intelligence sources indicated that most hostages were at a single location in Tehran. On June 27, the Air Force issued instructions to Lockheed-Georgia Company to begin preliminary engineering activity for a super STOL ver-

sion of the C-130. The C-130 was a logical choice. It routinely used JATO rockets to operate from landing strips in the Arctic and Antarctic. In 1963, a C-130 even demonstrated landings and take-offs from the aircraft carrier *U.S.S. Forrestal.*

The Lockheed-Georgia Company was to modify two (or possibly three) C-130 H aircraft into the YMC-130H SuperSTOL configuration. (Note, one source gave the designation as XFC-130H.) Initially, the aircraft were given the code name of Coronet Bat, but this changed at some point to Credible Sport. In July the Air Force acquired the task of providing a quick reaction force response team to manage the OSD program. Although the Air Force had overall project management, the Navy provided crucial assistance in the areas of rocket propulsion and arrested landing.

The Air Force and Lockheed handpicked the personnel to work this fast-paced program. Their task was to design and validate the STOL configuration and associated avionics needed to give the precise navigation and flight information the crew needed for such a surgical mission. The airframes were tested in September and October 1980, showing indeed that a government/industry team can respond quickly in an emergency.

Note the leading edge extensions for the horizontal and vertical tails. (Photo courtesy of Museum of Aviation)

By July 16, Lockheed determined that JATO power and arresting gear alone would not be sufficient. Eight days later, project planners had refined the proposal into a two-aircraft program, replacing the JATOs with rocket engines and making numerous aerodynamic and structural changes. By the end of July, the Air Force recommended that the $30 million project continue. Two aircraft on the production line were transferred to the program. (Note, one source reported three aircraft were pulled from operational service, tail numbers 74-1683, 74-1686, and 74-2065. The serial numbers would indicate that they certainly would have been in service for some years before being assigned to the program. It was confirmed, however, that 74-1686 is currently on static display at Warner Robins AFB in Georgia. Possibly, these aircraft were at the factory being remanufactured at the time.) The Air Force awarded a formal contract to Lockheed-Georgia on August 19. Lockheed promised to deliver the first aircraft for testing within a month, and to be mission ready within two months. This barely gave enough time before the scheduled presidential election on November 4.

The airframes were specially modified to take off and land in a very short distance, the landing spot being a soccer field with 10-meter-high obstacles at each end. The take off requirement was to obtain 90 feet of altitude within 600 feet from the starting point. The landing profile was not obtained, but probably was the reverse. Airframe modifications included, but were not limited to, the following modifications:

• Rocket engines to assist acceleration and braking, and to cushion the touchdown—four pairs of ASROC rockets mounted in fairings around the cockpit produced a total decelerating force of 80,000 pounds; eight Mk56 rocket engines, four to each side of the fuselage, pointing aft and downward, producing 180,000 pounds of thrust for four seconds, and 20,000 for another 20 seconds for acceleration; eight Shrike rocket engines, four on each side of the fuselage near the wheel wells that fired straight down to cushion the landing and aid the lift off; two ASROCs on the rear fuselage forward of the beaver tail, pointing down for pitch control and to prevent over rotation on take-off; and one Shrike near each wing tip, firing aft and down, for yaw control.

• Modifications to the T56 turboprop engines to give more power (for take-off, the engines and rockets together produced a total of 2 Gs).

• Horsal fins, which basically were horizontal stabilizer forward extensions.

• Dorsal fin fitted forward of the vertical tail to increase yaw stability at low speed.

• Flaps modified to a double slotted configuration.

• Extended chord ailerons.

• In-flight refueling pod from a C-141 installed above the cockpit on the fuselage centerline.

• Tail hook fitted forward of the rear cargo door for the planned carrier landing.

• Special passenger restraint system developed for the 150 passengers, up to 50 of whom, it was expected, might have suffered severe injuries. Also, a special pallet with seats for the Delta Force rescue team that could withstand up to 9 Gs upon impact.

• DC-130-type radome with a FLIR turret installed on the nose. The FLIR turret was slaved to the onboard computer and used in combination with the FLIR telemetry and other mission avionics to fire the forward pointing rockets 20 feet above the ground.

IBM developed the central computer. Texas Instruments provided the terrain-following/terrain-avoidance radar and FLIR/laser ranger system. Canadian Marconi's Doppler radar improved the accuracy of the navigation suite. These avionics would assist is performing the take off and landing, as well as ease low level navigation through Iran and help to find the landing spot. The various laboratories at Wright-Patterson AFB in Ohio helped integrate the system and test it.

The Credible Sport design philosophy was for a single point design for a single mission. Testing only explored a limited portion of the flying envelope. Because of the rapid pace needed, the program did not require the margins of safety normally required for peacetime operations, and the airframes were considered expendable. Everything had to work properly only for the flights in out of Tehran, and for the final safe landing with the rescued hostages.

A partially-modified C-130 was used to lower development risk. It was configured with the nose-mounted, wing-mounted, and tail-mounted rockets. These rockets would be fired automatically by the computer assisted "launch-sequence" system.

On September 18, the test-bed C-130 made its maiden fight. On September 20, it performed the first test of the decelerating rockets. This test determined that the T56 engines flamed out from ingestion of the rocket engines' exhaust, so the rockets were relocated to keep their exhaust out of the T56s' inlets. The best landing performance was obtained by firing the upper four rockets while still in the air, then the lower four immediately after touchdown. Tests also evaluated take off performance, and were completed in three weeks.

Three crews assembled to fly the actual mission, one each from Tactical Air Command, Pacific Air Forces, and United States Air Force in Europe. All were experienced operators of the MC-130E Combat Talon aircraft used for special operations. (Note, another gap in the information…it is not known how many crews might have been back-ups, as the number of aircraft that would perform the actual mission was not stated in any of the data.) As testing progressed, it became evident that the operational crews could not begin flying the aircraft before November 1, making the mission to Tehran before the election unlikely by this point.

The first fully-configured Credible Sport made its maiden flight on October 17. The enlarged ailerons had a flutter problem, but was fixed and flew again within two days. Following these tests Lockheed delivered the first of the two fully modified aircraft to the secure test area within the Eglin AFB complex to begin the test program. The first few flights at Eglin involved firing only two rockets, then a series of asymmetric burns to test the yaw correction rockets on the wings, which worked perfectly.

The first planned take off and landing under full operational conditions using all rockets began on October 29. On the take-off run, the nose gear lifted six feet into the air after less than a 10-foot ground roll. Within 165 feet, one and a half times its fuselage length, the Credible Sport was in the air. After a further 230 feet, it attained an altitude of 30 feet and a speed of 115 knots. The aircraft was stable and controllable throughout the take-off. Everything went properly, until just before touchdown. For reasons not clear, one of the crewmembers engaged the four ASROC retrorockets. The aircraft slowed too much while still 20 feet in the air and smashed into the runway, ripping off the right wing. The crash destroyed the aircraft, but the crew was not seriously injured.

Work on the second Credible Sport, which was near completion, continued at a feverish pace. On October 31, Tehran radio announced their intent to release the hostages, effectively ending the crisis. Preparations to support Operation Honey Badger continued into early 1981, but the pace that had characterized the military urgency of the program was gone. Iran released the hostages on January 20, 1981, concurrently as President Reagan was being sworn in as the new president.

Official USAF interest in Credible Sport subsided after the hostage release. Lockheed completed the second aircraft, but it did not fly before the program was terminated. It was transferred to Warner Robins AFB where it remained in service to further develop the avionics suite, but not the super STOL capability. The 1st Special Operations Wing asked that it be assigned to them for emergency rescue operations, but no such transfer ever took place. However, by the early 1980s, with the XV-15 proving the feasibility of the tilt rotor concept, the Air Force chose to develop the V-22 Osprey as the aircraft of choice for future surgical special operations missions. Subsequently, the surviving Credible Sport airframe was de-modified and given to the museum at Warner Robins AFB in Georgia in March of 1988. Many of the remnants of the external modifications are still apparent to the trained C-130 observer.

As a result of the data gathered from the Credible Sport program, Lockheed built a technology demonstrator called the High Technology Test Bed (HTTB for short) to further develop the STOL capabilities of the C-130 aircraft. They modified a company-owned L-100-20 (the civilian version of the C-130), registration number N130X, in 1984. Painted solid black, the HTTB sported many of the features developed for Credible Sport, but lacked the rocket engines. Unfortunately, the HTTB crashed at Dobbins AFB, GA, on February 3, 1993, during a high speed ground test run and was destroyed.